Plastic Inc.

Plastic Inc.

Big Oil, Big Money and the Plan to Trash our Future

Beth Gardiner

monoray

First published in Great Britain in 2026 by Monoray, an imprint of
Octopus Publishing Group Ltd
Carmelite House
50 Victoria Embankment
London EC4Y 0DZ
www.octopusbooks.co.uk

An Hachette UK Company
www.hachette.co.uk

The authorized representative in the EEA is Hachette Ireland, 8 Castlecourt Centre, Dublin 15,
D15 XTP3, Ireland (email: info@hbgi.ie)

Text copyright © Beth Gardiner 2026

Published by arrangement with Avery, an imprint of Penguin Publishing Group,
a division of Penguin Random House LLC, 1745 Broadway, New York, NY 10019.
First published in the United States in 2026.

All rights reserved. No part of this work may be reproduced or utilized in any form or by any
means, electronic or mechanical, including photocopying, recording or by any information storage
and retrieval system, without the prior written permission of the publisher.

Hardback ISBN: 978-1-80096-252-1
Trade paperback ISBN: 978-1-80096-253-8
eISBN: 978-1-80096-255-2

A CIP catalogue record for this book is available from the British Library.

Typeset in Adobe Garamond Pro.

Printed and bound in Great Britain.

1 3 5 7 9 10 8 6 4 2

This FSC® label means that materials used for the product have been responsibly sourced.

For my parents,
Ronnie and Barry Gardiner

The hierarchy of substances is forthwith abolished,
a single one will replace them all.

—FRENCH PHILOSOPHER ROLAND BARTHES, "PLASTIC," 1957

Oh great, we've upset the plastics industry?
This whole building is bankrolled by plastics.

—VICE PRESIDENT SELINA MEYER, AT THE US CAPITOL, IN *VEEP*

CONTENTS

Prologue . . . 1

CHAPTER 1
Plastic Dreams . . . 11

CHAPTER 2
"An Excellent Salesman" . . . 35

CHAPTER 3
America's Plastic Boom . . . 57

CHAPTER 4
"Guilt Eraser" . . . 83

CHAPTER 5
"A Fantastic Window of Opportunity" . . . 113

CHAPTER 6
Invisible Poisons . . . 133

CHAPTER 7
Wilson vs. Formosa . . . 163

CHAPTER 8
"They Want to Crush This Bug" . . . 187

CHAPTER 9
"No Silver Bullet" . . . 213

CHAPTER 10
"An Impatient Billionaire" . . . 229

CHAPTER 11
Shipping Plastic, Shifting Blame . . . 247

Epilogue . . . 273

Acknowledgments . . . 283

Notes . . . 287

Index . . . 335

Prologue

The headline felt astounding, unbelievable. My shock and confusion only grew as I read the article beneath, and by the time I'd finished, anger was bubbling up too. What had stopped me so short was the news that big fossil fuel companies were pouring billions of dollars into plans to make more plastic than ever in the years to come—even as so many people, worried about plastic's proliferation into every corner of modern life, were trying to use less. I was one of those concerned individuals—toting around a reusable water bottle, bringing my bags to the grocery store, and chiding myself a little every time I forgot them. It was painful to square all that with the reality I now understood: Some of the world's richest and most powerful corporations were pushing with all their might in the opposite direction. On the trajectory those companies are charting, I've since learned, plastic production—which has doubled in the past twenty years—will double again, or even triple, over the next few decades.

I could see that growth in my own life, where, despite my desire to reduce plastic's presence, there seemed always to be more of it—a flood of takeout containers, toothpaste tubes, and yogurt tubs; of packaging that encased every purchase arriving at my door and disposable utensils I would never use. I'd get a familiar twinge of guilt when I tossed it all away, then push the feeling aside, telling myself my family's contribution hardly made a difference to a global plastic heap already towering so high. There was nothing wrong with ordering from my favorite Indian place, or buying a new charger to replace the broken one, I'd think. And anyway, it wasn't my fault those ordinary decisions led so inevitably to the moment in which their baked-in wastefulness became visible.

That uncomfortable tangle of emotions is as good a window as any onto the unease that comes of being enmeshed in a way of life we know is endangering our collective future. Plastic's ubiquity distresses so many of us, I believe, because it's an inescapable, physical reminder of a frightening global mess—one we feel complicit in yet powerless to change. Which is why the revelation that the industry behind that mess planned to just go on adding to it—and, more than that, to increase the amount of plastic it churns out—lodged in my mind, eating at me.

In retrospect, it was the vast chasm between that galloping, unconstrained growth and my puny personal effort to consume less that started me on the reporting that would become this book. My worries about plastic, I realized, had been so focused on my own use of it, and on where it was ending up, that I'd barely considered where it was coming from. I'd seen the awful photos of turtles tangled in bags and beaches carpeted with litter. But while I was certainly of the opinion that retailers were using too much packaging, I hadn't stopped to wonder who was making all this plastic—and why.

I was aware, in the broadest way, that the material has its origins in oil and gas, but although I'm an environmental journalist, I didn't know much more than that. It felt, if I'm honest, like it came from nowhere, simply appearing when I needed it, and often when I didn't. It's a feature

PROLOGUE

of today's global economy that we don't have to think much about where the objects in our lives originate—the complex supply chains and hidden labor that bring us the things we want. That's even truer for plastic, a material so divorced from nature that the intuition able to tell me a wooden table was once a tree fails when I look at my toothbrush or a disposable fork.

That opacity suits its producers just fine. Plastic is a petrochemical, and those making it either are subsidiaries of giant fossil fuel companies—ExxonMobil Chemical, Shell Chemicals, and the like—or are deeply intertwined with them. While their product is omnipresent in our lives, they're happy to stay hidden in plain sight, minting money outside most people's awareness.

And while I didn't think I'd been harboring any illusions about Big Oil's civic-mindedness, what I've learned has been more shocking and unsettling than I'd imagined. Gradually, as the pieces fell into place, I could see it made sense. The plastic-drenched world we live in today didn't just happen. It was built by an industry that has slowly, steadily—and stealthily—drawn us into its web. These companies saw from the start that pushing ever more plastic into our lives would bring fat profits. So they came up with an endless series of new uses for it, inventing the lucrative idea of disposability, then pumping out a stream of throwaway packaging, cheap toys, and much more. They persuaded us to blame ourselves, and one another, for the resulting waste, to keep us from directing our gaze at their ever-climbing production. All the while, with tactics a city councilman who watched them in action decades ago said were worthy of Broadway—and with relentless, behind-the-scenes deployment of political muscle and campaign cash—they fought to stop anything that threatened their ability to sell more plastic, from bans on single-use items to a strong global plastic pollution treaty.

That project has been underway for decades. Today, there's a new, more sinister twist. Even oil and gas companies can see that if the world ever gets serious about combating climate change, our previously insatiable appetite for their fuels will diminish. Making more plastic has become their plan for hedging against that possibility—a way to keep profits flowing even if electric vehicles and renewable energy begin, at long last, to dent oil and gas demand. While they're doing everything in their power to put the brakes on a global shift toward clean energy—and despite the fires and the floods and the heat and the droughts, it sometimes looks like they're winning—they are also smart enough to plan for it. An industry whose greed has brought us to the brink of planetary catastrophe now sees plastic as the answer to the threat climate action poses to its bottom line.

Indeed, with petrochemicals poised to become the largest single driver of oil-demand growth, making ever more plastic is now a central pillar in Big Oil's dream of continuing to drill for as long as it possibly can—a way to indefinitely perpetuate an epically destructive business model. And plastic's role in incentivizing continued fossil fuel extraction is just one piece of its climate harms. While I'd long thought of these two ecological catastrophes as distinct, the reality is that plastic worsens warming at every stage of its life, from its hugely polluting production to the incinerators that burn much of our waste. "If plastic were a country," one analysis said, "it would be the world's fifth-largest greenhouse gas emitter." That means facing down plastic's proliferation is a necessary part of the fight for a livable climate.

It's clear to me now how much coverage of plastic and its attendant harms, and therefore the public conversation about it, is blinkered by the same myopia I'd suffered from. It makes sense. We focus on what we can see: either the symptoms of this problem—the plastic filling oceans and piling up in landfills—or our own actions, and those of the people around us. The decades of corporate and political decisions that have determined what choices are available to us are less obvious to the naked eye. As anyone who's tried to use less plastic knows, it's hard to avoid.

PROLOGUE

Often, there's no alternative, something we need or want comes wrapped in it, or avoiding it would be expensive or require a great deal more effort.

To be sure, there is value in the small steps that can shrink plastic's footprint in our own lives. My own family, for example, recently began getting milk delivered in glass bottles by a company that collects and reuses them. Such actions can shift markets, seed larger changes, and help show what a different world might look like. But in concentrating on them too exclusively, we can miss an important reality: It is the actions of almost incomprehensibly wealthy corporations with tremendous political power, not our everyday choices as individuals, that are driving our spiraling global crises.

So I decided to train my gaze on that big picture rather than my own little piece of it. Indeed, I'd soon see that plastic is where the "individual action" frame collapses under the sheer weight of corporate cash. We may be entangled in this ugly mess, but I've learned we are not its main actors. In the pages that follow, I'll introduce you to some who are, as we explore the past, present, and future of an industry whose guiding force has always been its voracious hunger for more—more growth, more markets, more profit. More plastic.

We can see the change all around us, advancing every year: bars of soap replaced with pumps of liquid, coffee filters giving way to plastic pods, disposable cups and plates used even while eating in at restaurants. We never asked for all that—for fruit and vegetables to be swathed in plastic, or unwanted utensils stuffed in the take-out bag, headed straight for the trash. It happens without our say-so.

Plastics' spread into every corner of our lives is also taking an invisible toll on our health. The chemicals that leach from them have been linked to heart disease, cancer, fertility problems, and even neurodevelopmental issues such as autism and attention deficit hyperactivity disorder. Tiny microplastic particles now suffuse the air we breathe, the food we eat, and the water we drink, and have been detected in human brains, hearts, bone marrow, placentas, and breast milk.

I sometimes refer to the "plastic industry" in the singular, but it is in fact a set of interconnected, interdependent industries. Fossil fuel companies providing ingredients. Petrochemical firms turning those oil- and gas-derived substances into raw plastic. Manufacturers shaping it into everything from hair clips and car parts to bags, pouches, and bottles. The huge conglomerates that buy and use such packaging: retailers like Amazon and Starbucks, food and drink makers such as Coca-Cola and Nestlé, and consumer products giants Procter & Gamble (whose dozens of brands include Gillette, Pantene, and Pampers) and Unilever (maker of Hellmann's, Dove, and many more). Their interests overlap at times and diverge at others. But, with hundreds of billions of dollars at stake, they've all worked to make—and keep—plastic central to our lives.

Indeed, its presence is even more wide ranging than I'd once imagined— and given this material's innumerable varieties, the plural more accurately captures reality even in its very name. Plastics, for instance, line the insides of take-out coffee cups and other paper and cardboard containers. They comprise the bulk of most wet wipes, diapers, and sanitary pads. Tires' synthetic rubber is a plastic, as is artificial grass. It's in our clothing and shoes, and tiny, invisible beads of it make lipstick, eyeliner, and mascara smooth and spreadable. Mattresses and furniture stuffing, curtains and carpets—all are often made, in large part, from one kind of plastic or another. Needless to say, it adds up to an enormous mountain of money—more than $760 billion annually, by one estimate, and climbing by 4 percent a year.

Of course, this is not a simple story of an evil material. Plastic has many valuable, even essential, uses. I wouldn't want to live in a world without the medical equipment it makes possible. Our food supply chains are predicated upon its ubiquity and affordability, and it's hard to imagine a modern supermarket without it. Industry representatives like

PROLOGUE

to note that plastic makes cars lighter, and therefore more fuel-efficient, and that it's a key part of solar panels and wind turbines. They're not wrong. But plastic's defenders talk up such uses to distract us from the huge wastefulness they've created—to confuse us into thinking that if we want sterile gloves and syringes, we must also accept a tide of useless junk.

As I began to reframe this story through a lens of corporate and political—rather than personal—responsibility, I kept returning to one question. How could they do this? With plastic already infusing our lives and blanketing the natural world, why would fossil fuel and petrochemical companies keep pushing to sell us even more? My angry disbelief alternated, and intermingled, with an understanding that these industries have always put profit above all else. Still, I wanted to understand how companies had landed on this strategy, how long they'd been pressing it, whether they were likely to succeed, and—crucially—what might stop them. I wondered, too, where to find the line distinguishing plastic's valuable uses from its unnecessary ones—and how we could keep what matters while avoiding the harm from so much wasteful excess.

The search for answers would take me from Arizona's cactus-dotted mountains and Houston's muggy bayous to the narrow cobblestoned streets of Antwerp and the glitz and bling of Dubai. And from virtual rabbit holes lined with decades-old conference transcripts to those made from oil analysts' predictions and companies' own—sometimes euphemistic, but always telling—words. I'll show you what I found on that journey, as we take a close-up look at the petrochemical companies turning molecules buried deep underground into plastic—and dollars.

We'll see how this vast, rapacious industry was born from a drive to monetize what was worthless, and how it grew along with modern consumerism. While much of this history is set in the United States, it's shaped lives and economies far beyond America too. Companies quickly exported approaches they honed in the US, from eliminating reusable packaging and the systems that supported it to weaponizing lies about recycling—persuading us it was a panacea for all the problems plastic might pose when they knew it would never work.

On the sunbaked Texas coast, I'll introduce you to a former shrimp-boat captain who waged a groundbreaking legal battle against the huge conglomerate spilling plastic pellets into the bay she loved and to the buttoned-down Taiwanese tycoon who built the company. We'll see how fracking has unleashed a huge boom in US plastic production, and how plastic, in turn, is fueling more fracking—despite industry and its supporters having sold this harmful form of drilling to the public as a way to provide energy. And as we follow the global tentacles of an industry whose web stretches from Saudi Arabia to China and far beyond, we'll meet a lanky British billionaire whose penchant for taking big risks is matched by a knack for making them pay off.

One thing I've often heard when telling people about this book is that it sounds depressing. I get that. Sometimes I feel it too. The scale of the money and power aimed at drowning us in plastic can seem overwhelming. Mostly, though, I find this industry's untrammeled greed and utter disregard for the consequences infuriating instead. And in an odd way, I think looking it in the face is empowering. In this era of misinformation, viral conspiracy theories, and a loss of trust in shared realities, it can feel naive to hold on to a faith in the power of facts to bring change. But I've built my professional life, and much of my personal worldview too, on the belief that they matter.

Environmental writers are frequently asked—by our readers and editors, our friends and families—to offer some hope. I've always had mixed feelings about that. It's not really my job. Journalism is about telling the truth, as best we can, not making people feel better. On the other hand, it's neither fair nor helpful to leave readers in despair. In the end, I reconcile these conflicting ideas with the notion that hope starts with clear-eyed understanding. After all, you have to diagnose a disease before you can treat it. If we want to check plastic's out-of-control proliferation, we must first understand what—and who—is causing it. When we do, we'll realize we can stop blaming ourselves, and be freed to move away from the small-bore solutions that feel so futile and toward shifts that match the scale of the problem.

With apologies for the spoiler, one thing I've learned in reporting *Plastic Inc.* is that it's not just corporate responsibility but corporate scandal that offers the most useful lens for this story. And the stakes are even higher than with the more familiar depredations of industries like tobacco and opioids. Indeed, it is Big Oil's own defining scandal—first pushing to discredit climate science, now paying lip service to sustainability while trying to delay action—that offers the best template for understanding plastic producers' plans and the threat they pose not just to the health of individuals but to the well-being of humanity as a whole on a livable planet.

Plastic is woven into our economy and our lives, and scaling back its presence won't be easy. It's possible, though, and there are examples of progress all around. Even with everything I know, I'll keep carrying my canvas bags, my reusable water bottle and coffee cup. But real change, I believe, must start with a hard look at those whose insatiable hunger for more has long been the biggest obstacle. While plastic itself is easy to see, the industry bringing it into the world is invisible to most of us. This book aims to shine a light on it.

CHAPTER 1

Plastic Dreams

1951—Businessman Earl Tupper hires a marketer to plan in-home sales parties for his Tupperware containers.
Three million metric tons of plastic produced.

" **W**axy solid found in reaction tube." Those were the words Reginald Gibson scrawled, in spiky, almost illegible script, when he arrived at his lab in the northern English county of Cheshire one morning in March 1933. Along with fellow researcher Eric Fawcett, he'd been experimenting with a chemical called ethylene, which is derived from oil and gas. They'd been combining it with other substances under high pressures, but when they tried to repeat the reaction, it only occasionally produced that waxy solid. More often they got a loud bang and a puff of smoke. Fawcett and Gibson believed they were onto something significant, but their bosses at Imperial Chemical Industries worried another explosion might do real damage. So they stopped the project.

A couple of years later, the company had developed better equipment, and a research director named Michael Perrin began running the experiment again, with two colleagues. On the very first try, they got a pile of white powder that soon hardened into the same stuff Fawcett and Gibson had produced. "It was a fluke," a lab assistant later recalled. But it gave them the hope they needed to press on. Eventually, the team figured out what Fawcett and Gibson hadn't understood: Oxygen had sometimes leaked into their equipment, and it provided the chemical key to creating the waxy substance.

With that insight, the scientists could synthesize the new material at will, and they made enough of it to confirm what Fawcett had suspected. The waxy solid was polyethylene—an almost never-before-seen material composed of extraordinarily long chains of fused-together ethylene molecules. The British called it polythene, but by any name, it had some intriguing properties. When heated, it could be molded into shapes. Mixed with a chemical called xylene, it dissolved into a gel that could be spread thin to dry into a flexible film. And it did not conduct electricity. By 1937, Imperial Chemical Industries had secured a patent and was talking it up to potential customers. The following year, it began taking orders for batches to insulate underwater telecommunications lines. On September 1, 1939, the company fired up a new reactor consisting of two fifty-liter vessels that could together produce one hundred tons of polyethylene a year. It was the same day Germany invaded Poland, the blitzkrieg that began World War II. The timing, if painful, was propitious.

Today, polyethylene is the world's most common plastic. But before it became food wrap, shopping bags, bottles, and coating for paper coffee cups, it gave the Allies a critical advantage. In the early months of war, British scientists were developing a sophisticated new form of radar, but without high-quality insulation, the signals running through its wires would escape before they could be read. The right material would enable radar sets to be made small and light enough to be loaded onto planes without sacrificing their effectiveness. Polyethylene fit the bill perfectly, making an "almost insoluble" problem "comfortably manageable," one of radar's pioneers later said. As Londoners endured the Blitz, Royal Air Force pilots in 1941 began using their new gear to locate and shoot down Nazi bombers. By mid-1943, with even better radar—thanks to polyethylene-coated cables—Allied planes could spot German submarines. In the month before the new technology was introduced, the Allies lost fourteen ships to Nazi U-boats, whose attacks had imperiled critical supply lines from the United States to Britain. For nearly five months after, they lost none. "Enemy aircraft have been equipped with a new location apparatus" enabling them to detect subs even at night, a German admiral reported to Hitler. "The tide had turned," one historian wrote, and Germany never regained its naval supremacy. Lacking polyethylene, the Nazis had to use bulkier insulation that made their radar both less effective and less portable. It was an early hint of the ways this extraordinary new material—lightweight, versatile, and (eventually) inexpensive—would change the world.

Polyethylene was not the first plastic. In the mid-1800s, consumer demand for ivory—used in piano keys, chess pieces, knife handles, and more—was driving rampant slaughter of elephants. Tusks good enough to make billiard balls were particularly expensive, so in the 1860s a New York company that had helped popularize the game in the United States offered $10,000 to anyone who could come up with an alternative material. John Wesley Hyatt, then running a print shop upstate in Albany, started compressing scraps of cloth, paper, and wood into spheres he covered with shellac. Eventually, he added collodion—a gel printers used to protect skin from hot lead type—as a coating over spheres of wood and pressed cotton. The only trouble was the minor explosion when two banged together. "Every time the balls collided, every man in the room pulled a gun," reported a saloonkeeper testing them out.

Hyatt kept tinkering and eventually tried camphor, distilled from the sap of laurel trees, as a solvent. Combining it with a cotton paste treated with acid, and applying heat and high pressure, he created a malleable, moldable material and in 1879 patented it with the name celluloid. Soon it was being used to make not just billiard balls but dentures, combs, and, most famously, film for still and moving pictures. If such items sometimes tended to explode or catch fire—or, in the case of dentures, to taste of camphor—they nonetheless offered huge advantages over those made from naturally occurring materials like wood, rubber, ivory, tortoiseshell, silk (from cocoons spun by silkworms), and shellac (secreted by the Southeast Asian lac beetle). They were cheaper, available in potentially endless quantities, easy to mold into nearly any shape desired, and could be produced in an array of colors and textures. One of the main selling points feels sharply ironic now: By breaking the link between manufactured goods and the natural world, celluloid and other early plastics meant it would "no longer be necessary to ransack the earth" to make the things humans wanted, as Hyatt's new company declared.

More new plastics—the word comes from the Greek *plassein*, meaning "to mold or shape"—soon followed. In 1907, Belgian immigrant Leo Baekeland synthesized another important one, Bakelite, in New York, from the chemicals formaldehyde and phenol, the latter a waste product from coal. Chemically, these new materials were known as polymers. The word means "of many parts," and refers to the long chemical chains that make them up. That length gives them their flexibility and strength. Polymers can occur naturally, like cellulose in plants' cell walls—the stuff Hyatt's celluloid took from cotton. Purely synthetic ones—such as Bakelite and its successors—are typically longer and more complex. Exactly what's in the chain and how its molecules are bound together determines the plastic's type and characteristics.

Crude oil and gas—containing chains of carbon and hydrogen that could, with high pressures and temperatures, be pried apart and recombined into even longer configurations—provided a rich set of raw materials. By the 1920s, and into the '30s, scientists across Europe and North America—Imperial Chemical's Eric Fawcett and Reginald Gibson among them—were experimenting with chemicals derived from the processing of fossil fuels. Materials that would go on to become ubiquitous began to make their first appearances—polystyrene; polyvinyl chloride, or PVC; polymethyl methacrylate, marketed as Plexiglas (Perspex in Britain). Not everyone grasped their significance. Shortly after Fawcett and Gibson's lab accident, their company created a Plastics Division. "By the way, what are plastics?" its first director asked upon accepting the job. "God knows," one of his new bosses answered. "But you might show some enthusiasm." Soon the new materials' potential became impossible to miss, as they swiftly entered daily life in items like telephones, radios, furniture, and car parts. In 1938, DuPont patented the synthetic fiber nylon. Researchers had come up with two potential recipes for a silk substitute. Against the wishes of the project's lead scientist, the company opted for the one based on benzene, a chemical that—in addition to being a potent carcinogen—was readily available as a by-product of coal processing (today it more often comes from oil and gas). Like polyethylene, nylon would prove essential in the war, used for gear including parachutes and rope.

The creators of the earliest plastics had often sought to fill a specific need—as John Wesley Hyatt hoped to do in the billiards company's contest. But many in the new cohort of chemical makers came at their work from the opposite direction, seeing what materials they could create and figuring they'd find useful applications later. "The inventor is now in advance of the wants of his time," one of the founders of Fawcett and Gibson's lab had said, in a speech summing up its ethos. "He may even create new wants."

That approach foreshadowed one of plastic's most distinctive properties: its power to reverse the normal relationship between supply and demand. For decades now, the material's very existence—and its low price—have propelled the relentless invention of new uses for it. The inversion of supply and demand goes even deeper than that, driving plastic's production and its astonishing ubiquity. It's a dynamic that stems directly from the material's relationship to fossil fuels. An anecdote that's probably apocryphal nonetheless illustrates the forces that have shaped plastic's creation and subsequent growth. According to legend, when John D. Rockefeller—whose Standard Oil eventually split into companies that became ExxonMobil, Chevron, and BP—noticed flames lapping up from one of his refineries' smokestacks, he asked what was burning. It was ethylene, created as a by-product of the refining process. "I don't believe in wasting anything," he supposedly snapped. "Figure out something to do with it." The story neatly sums up petrochemical makers' ethos: Find a way to wring money from every molecule coming out of the ground.

That was how Fawcett and Gibson ended up experimenting with ethylene—and it was no coincidence that Fawcett had worked for a time at the American Petroleum Institute, an alliance of the biggest US oil and gas companies. Polyethylene fulfilled Rockefeller's (apocryphal) dream more spectacularly—and more profitably—than perhaps even he could have imagined. The petrochemical companies that grew through the middle of the twentieth century, turning fossil fuel–derived ingredients into not just plastics but also fertilizers and the raw materials for detergents, cosmetics, lotions, paints, and more, were often subsidiaries or offshoots of the very firms producing the oil and gas. Eventually, the biggest makers of plastic and other petrochemicals would include ExxonMobil Chemical, Shell Chemicals, China's Sinopec, and the Saudi Basic Industries Corporation, or SABIC, largely owned by the oil behemoth Saudi Aramco. Even those that were independent entities—big names like DuPont, Dow, Union Carbide, and Germany's BASF—were deeply intertwined with the fossil fuel industry, bound together in mutually lucrative interdependence. Often, they built their huge chemical plants beside refineries for easy access to the stuff left over after fuels such as gasoline, diesel, and kerosene were siphoned off. Those raw materials' origins as unwanted by-products are the reason for plastic's low cost. Instead of waste that needed to be disposed of, gases like ethylene became an opportunity for more profit. And the expense of constructing petrochemical plants meant they'd need to pump out vast quantities of product to be worthwhile. Like plastic itself, the upending of supply and demand that drives its proliferation was born out of oil and gas.

AFTER WORLD WAR II, IT WAS CLEAR THERE WOULD BE ENDLESS PEACETIME applications for the materials that had been so useful militarily. Researchers got to work tweaking existing plastics to create new formulations, each with distinctive characteristics—varying degrees of hard and soft, rigid and flexible, opaque and transparent. Polyethylene alone got an alphabet soup of iterations, eventually including—but not limited to—HDPE (high-density polyethylene), MDPE (medium-density), LDPE (low-density), VLDPE (very low-density), LLDPE (linear low-density), and cross-linked (PEX). The tongue-twisting names of the multitude of new plastics being synthesized—polypropylene, polystyrene, polyurethane, acrylonitrile, among innumerable others—belie their ubiquity, and familiarity, in our everyday lives. While the word *polymer* refers to the "many parts" of such materials' molecular structure, it also neatly captures the reality that plastic is not one thing, but many—thousands of distinct substances subsumed into one word.

As this bevy of plastics became available, marketing teams brainstormed ways to get them into people's hands. Plastic producers saw early on that reaching those designing and making products would be just as important as influencing shoppers—"selling the manufacturer," as an editor of the trade magazine *Modern Plastics* put it, on the idea that plastic was not only a superior material but one that could make goods more appealing and get shoppers reaching for their wallets.

Of course, the public had to be wooed too. Days before the atomic bombings of Hiroshima and Nagasaki, J. W. McCoy, a vice president at chemical giant DuPont, a leading plastic maker, said in a meeting about the coming transition to a postwar economy that a backlog of unmet demand would keep manufacturers busy for a time. But "a satisfied people is a stagnant people," he warned, so it would be necessary "to see to it that Americans are never satisfied." *Harper's Magazine* editor Frederick Allen had put his finger on the problem years earlier, after a brief economic downturn at the start of the 1920s got manufacturers worried about what might happen if people didn't want all the things they'd begun churning out. Unless consumers "could be persuaded to buy and buy lavishly," Allen wrote, "the whole stream of six-cylinder cars, [radios], cigarettes, rouge compacts, and electric ice-boxes would be dammed up at its outlets." The imperative was clear: If companies selling such goods were to succeed, workers for whom they had until recently been out of reach had to be educated in "skills of consumption," as one corporate adviser wrote. "The future of business," an advertising journal opined, "lay in its ability to manufacture consumers as well as products."

With plastic throwing productivity into higher gear, the task was becoming even more urgent. Suddenly, a manufacturer could "make a list of the properties which he would like to find embodied in a new material" and "custom-build that material as he never could before in all history," *Harper's* explained. And factories' own evolving technologies boosted output further. "In the thirty seconds it took you to read this far, a modern injection molding machine turned out sixteen combs from a single mold," a piece by chemical maker Monsanto proclaimed. Just learning to buy more was no longer enough. To absorb all the goods rolling off production lines, Americans would have to "make consumption our way of life," one marketing consultant explained, to "seek our spiritual satisfactions, our ego satisfactions, in consumption."

Advertising—or "want creation," as the economist John Kenneth Galbraith called it—would be key to making that happen. A billion-dollar business in 1920, it was worth $10 billion by 1956, its reach and power growing with the advent of television. Edward Bernays, a Vienna-born immigrant now commonly described as the father of public relations, had set out the tenets that would guide the field's practitioners. "To assure itself the continuous demand which alone will make its costly plant profitable," a business "cannot afford to wait until the public asks for its product; it must maintain constant touch, through advertising," he explained in his classic 1928 book *Propaganda*. Bernays was a nephew of Sigmund Freud, and to achieve "intelligent manipulation of the organized habits and opinions of the masses," he weaponized his uncle's insights. An individual might want something "not for its intrinsic worth or usefulness," he observed, but "because he has unconsciously come to see in it a symbol of something else, the desire for which he is ashamed to admit to himself." With an understanding of such buried motivations, Bernays advised, "the successful propagandist" can "swing emotional currents so as to make for purchaser demand." For Lucky Strike, he framed cigarettes as "torches of freedom" that empowered women. Fear was also effective. To sell Dixie cups, Bernays formed a Committee for the Study and Promotion of the Sanitary Dispensing of Food and Drink to suggest reusable cups were unhygienic.

As plastic makers sought to cement their product's place in postwar life, they also helped engineer coverage touting the amazing changes it would bring. When *House Beautiful* began working on a special issue about the miraculous new material in 1947, the chief publicist at the Society of the Plastics Industry, the companies' main trade association, served as consultant. Not only did he help answer writers' and editors' questions; he also suggested story ideas and rewrote copy. It paid off. "Plastics . . . a Way to a Better More Carefree Life," the cover proclaimed. Inside, the magazine told readers that "plastics are here to free you from drudgery" and held the potential to "improve your life a thousandfold."

After the long years of depression and war, consumers were all in. They snapped up the new products, and new plastic versions of old ones, crowding the shelves of supermarkets and department stores—toothbrushes, toilet seats, ashtrays, furniture, luggage, appliances, and much more. The shift that had begun in the 1930s was complete. Plastic—the foundational material of modern consumerism—was at the heart of a new age of plenty, bringing once-unimaginable convenience to everyday life. A vision the chemical conglomerate Monsanto had summed up in a 1947 headline in its company magazine was becoming reality: "Doubling—Tripling—Expanding: That's Plastics." It was an apt description of not just the industry's dreams for plastic but its hopes for Americans' appetite for new goods more broadly—and for the profits the convergence of those two forces would bring.

AS IT WOULD AGAIN AND AGAIN IN THE YEARS TO COME, PLASTIC'S AVAILability and affordability created the incentive for enterprising businesspeople to devise uses for it. In the early 1950s, with eight different companies building polyethylene plants, that material's price dropped below twenty-five cents a kilogram, unleashing a flood of new goods, whose popularity then juiced production further. The baby boom generation's passage through childhood offered particularly fruitful opportunity. During the war, General Electric had assigned an engineering team to respond to a government plea for synthetic rubber. The stretchy stuff they came up with wasn't militarily useful, and languished in GE's lab until one of its designers pulled a wad out at a 1949 party. A toy store owner and an ad writer asked if they could list it in their catalog, but even they were surprised by its popularity. It was the adman, Paul Hodgson, who bought a ton from GE, for $147, and packed it into egg-shaped containers (themselves also plastic). Silly Putty became the fastest-selling toy in US history.

A similar dynamic was on display when Phillips Petroleum found itself stuck with an excess of off-spec plastic a few years later. The company was developing a new high-density form of polyethylene, but early batches were brittle and cracked easily. In 1957, when the Hula-Hoop became a culture-defining craze, Wham-O, its Californian creator, found its original plastic supplier couldn't keep up with demand. The toy maker bought Phillips's unwanted stockpiles, then ordered even more. When hula-hooping's popularity faded, Wham-O moved on to Frisbees. If insulating radar had been an indication of plastic's beneficial, even life-saving, possibilities, the Hula-Hoop was an early hint of the ways this material would be put to far less essential use too.

The fad's quick ebb illustrated another emerging dynamic. Even as the explosion of new goods elevated buying things to a central place in American life, it seeded the idea that many of those objects were of so little value "as to encourage their displacement, their disposal, their quick and total consumption," historian Jeffrey Meikle wrote in 1995's *American Plastic*. Plastic was the perfect match for that ethos, he observed. Its low price—"because it was derived from an endless supply of petroleum"—was key. "It was, finally, so lightweight and in some forms so insubstantial as to be discarded without a second thought."

Like the coming and going of fads, changing fashions was another way to gin up sales. Foster Grant, a plastic-molding firm that had begun making sunglasses, found frequent style updates could prompt customers to buy multiple pairs. Designing goods to wear out quickly so they would, as one car-part manufacturer explained, "get to the junk pile faster" worked too.

Other products were more explicitly intended to be thrown away as soon as they were used. In 1961, Procter & Gamble introduced Pampers, the first disposable diapers, freeing parents—mainly mothers—from a burdensome chore, but creating a plastic mountain that would eventually clog landfills with hundreds of billions of diapers. Throwaway cups, plates, and utensils appeared on store shelves, along with pens and cigarette lighters that—unlike older, refillable versions—had to be tossed once empty, and toys that broke so quickly they were practically disposable too. The trade magazine *Modern Plastics* set out what that meant for industry: "The factor of disposability," it explained, was "an important key to continuing [sales] volume." Even if consumers had to be nudged toward this new relationship with objects, it wouldn't be hard. In the late 1950s, the journal noted, Americans sometimes saved single-use items, but "it is only a matter of time until the public accepts the plastics cups" as being "completely discardable."

CARROLL MUFFETT CALLS HIMSELF "A RESEARCH GEEK AND A RESEARCH junkie." An environmental lawyer with a round, boyish face and the hint of a Kentucky lilt in his voice, he has spent years digging through archives and unearthing old documents in an effort to understand the inner workings of the fossil fuel industry. The question driving him when he began, around 2008, was simple: How early had oil companies realized their product would cause dangerous disruption to the climate? Ordering decades-old reports, books, and magazines off Amazon and eBay or from academic libraries, he uncovered evidence that helped shape public understanding of the scandal that came to be called "Exxon Knew." The truth Muffett helped uncover—along with journalists from *Inside Climate News* and the *Los Angeles Times*—was that scientists working for the biggest oil and gas companies, including but not limited to ExxonMobil, clearly anticipated, at least as early as the 1960s, that the unchecked burning of fossil fuels would dangerously warm the atmosphere, causing heat, drought, storms, and floods that could imperil humanity's future. And not only did they keep that understanding to themselves; they aggressively pushed to discredit anyone speaking publicly about it. By the 2010s, thanks in part to Naomi Oreskes and Erik M. Conway's landmark book, *Merchants of Doubt*, it was clear, too, that in their relentless campaign to prevent, or at least delay, climate action, oil companies were using the same tactics the tobacco industry had deployed to continue selling cigarettes in the face of mounting evidence of their dangers—funding scientists to produce industry-friendly "evidence" and attacking independent researchers who reached unwelcome conclusions. Above all, both industries weaponized the notion of uncertainty, aggressively peddling the idea that a question of near-universal scientific consensus remained unsettled, open to never-ending debate. The strategy, as *Merchants of Doubt* quoted one tobacco executive declaring, boiled down to this: "Doubt is our product."

Around 2015, Muffett, who heads the Washington-based Center for International Environmental Law, found himself wondering about the relationship between the two industries whose tactics were so often likened. As a result of dozens of states' lawsuits against tobacco interests, fourteen million industry documents had been deposited in an archive held by the University of California, San Francisco, much of it available online. To begin teasing out connections, Muffett decided to look for instances of tobacco representatives sitting on oil company boards, and vice versa. "I'm not making this up," he tells me on a Zoom call. "Within the first fifteen minutes of searching," he found numerous examples, including the chair of British American Tobacco, one of the world's biggest cigarette companies, who sat on Exxon's board in the 1970s. "I was like, 'Well, so that's done. What else can we find here?'" He started delving into what such ties between the industries might have meant. His group didn't have funding for this particular line of research, so he did most of his trawling at night and on weekends and holidays. The archive was a treasure trove, but "it really requires brute-force searching," he explains, sometimes "through unorganized folders with no names, for hours on end."

Before long, he saw that a desire to cast doubt on solid science wasn't the only thing the tobacco and oil industries had in common. They shared overlapping commercial interests too—for decades, gas stations had been among the most important retail outlets for cigarette sales, and cigarettes were the stations' biggest-selling item after gasoline. "The deeper I looked, the deeper the connections became," he recalls. Then he stumbled on something even more unexpected: documents showing oil companies such as Shell and Esso (part of today's ExxonMobil) had helped tobacco firms including R. J. Reynolds and Philip Morris analyze the makeup of cigarette smoke in the 1950s and '60s. Oil firms were well positioned for such work because it required the same technology and expertise they used to study the air pollution their fuels caused. Sometimes they even did the cigarette testing themselves, with tobacco industry funds. The discovery "was an 'Oh wow!' moment," Muffett recalls. It felt like "an extraordinary thing to know, that these two denial efforts were linked at this really seminal moment."

It led him to another question: What had come of the testing? The answer was right there in black and white, and it's where plastic comes into the story. No one seemed to consider warning smokers about what they were puffing. Instead, oil companies wondered if their petrochemical arms could start producing cigarette filters. Tobacco firms had been experimenting for years with a variety of filter materials, including cellulose acetate, a plastic made—like John Wesley Hyatt's celluloid—from the cellulose in plants' cell walls (in this case, typically wood pulp). Carroll Muffett found contracts and other documents showing oil companies' cigarette smoke analysis had quickly morphed into work designing filters from polypropylene, a fossil fuel–derived plastic. A 1967 memo by a Philip Morris employee details his conversation with a consultant who "informed me that Shell has several cigaret [sic] filters and wondered concerning our possible interest." A year later, a researcher at tobacco firm Liggett & Myers met representatives from Esso and Enjay Chemical, both arms of Exxon, who asked whether his company would be interested in polypropylene filters. "They had with them some samples" and mentioned "that they had approached two other tobacco companies," he wrote. Muffett and his colleagues began searching for patents and found both Shell and Esso had gotten them for polypropylene filters. At first, Muffett tells me, "I was absolutely staggered" by the whole thing. But he knew a lot about fossil fuel companies by that point, and once he got past his initial shock, their response made perfect sense. "What they saw was another market opportunity for their plastics," he says. "And they seized it."

In the end, it didn't really work out. Polypropylene filters sold for a while, and efforts to market them continued into the 1990s, as did at least one joint venture between Exxon and R. J. Reynolds. But their problems, as handwritten notes from one Exxon meeting recorded, included a taste smokers found unfamiliar. Today, most filters are cellulose acetate, one of the few plastics not derived from fossil fuels. An estimated 4.5 trillion butts are improperly disposed of every year, making cigarette filters the world's most common form of plastic pollution. They leach toxic chemicals that can be deadly to sea creatures, and each one can degrade into thousands of microplastic fragments.

And unfortunately, filters don't actually make smoking less hazardous—indeed, they can even make it worse, by adding synthetic fibers and particles to the toxic brew, and prompting smokers to change the way they inhale. One expert calls them the "deadliest fraud in the history of human civilization." But they were an effective marketing tool. A confidential 1966 Philip Morris memo on what the company called "health cigarettes"—jaw-dropping in its bald-faced manipulation—said "the illusion of filtration is as important as the fact of filtration," and referred to one filter type as "an effective advertising gimmick." To Muffett, the cigarette filter is "emblematic of a product that serves no purpose. It is counterproductive throughout its entire life. It only does harm." At the same time, it's just one example—particularly pernicious, to be sure—of oil companies' relentless focus on ginning up demand for plastics. Seeing how eager Exxon and Shell were to create new markets for polypropylene,

Muffett says, "really emphasized for us that plastics are, at the end of the day, fossil fuels in another form."

Another thing he took from his research was that while it's become common to think of Big Oil's climate denial as drawing on a Big Tobacco playbook, the borrowing more likely happened the other way around. As cigarette companies in the 1950s began seeking to defuse worry caused by evidence of smoking's dangers, the industry's powerful public relations firm, Hill & Knowlton, deployed on their behalf tactics it had honed helping the oil industry, which back then was pressing to undermine the science on air pollution's dangers, including the terrible toll of lead in gasoline. Muffett found multiple instances of tobacco companies hiring scientists and organizations who'd done such work for fossil fuels. Over and over again, his environmental law group said, the archives revealed how tobacco companies and their trade group, the Tobacco Institute, "turned to the oil industry for advice, for models, and, ultimately, for people to carry out their campaign of denial and deception," publicists and researchers who "worked first for oil, then for tobacco." The denial playbook, Muffett's team concluded, "originated not with tobacco—as long assumed—but with the oil industry." While efforts to question the link between smoking and cancer have run out of road, fossil fuel firms' push to stop serious climate action is still very much alive. The industry's continued focus on plastic is at once a symptom of and a strategically central plank in that denial.

TODAY, PLASTIC IS EVERYWHERE, A DEFINING, INESCAPABLE PART OF modern life. Among its many applications, it's become a key construction material, used in pipes, cables, insulation, windows, and doors. It's in cars and planes, composing all or part of their seats, tires, paneling, fenders—even the structure of aircraft fuselages. Clothing is another huge plastic user: Synthetic fibers such as polyester and nylon account for nearly 70 percent of textiles, sometimes mimicking materials like cotton, wool, or linen, and sometimes more obviously artificial. Most sneakers are made almost entirely from plastic—insoles of ethylene vinyl acetate, uppers in nylon mesh, polyurethane cushioning, synthetic rubber soles, polyester laces. Electronic goods—laptops, phones, TVs—are composed largely of highly engineered, specialized plastic. It plays a key role in modern medicine—gloves, syringes, tubing, IV bags, not to mention artificial joints, limbs, and heart valves. The list, of course, goes on and on—appliances, eyeglasses, synthetic grass, the frames and fillings and fibers of our furniture. Many uses feel essential, but others are beyond absurd: On a beach vacation a few summers ago, I was flabbergasted to find a clear plastic drip catcher on the bottom of my ice cream cone, marring what is otherwise a perfect bit of edible packaging.

Indeed, half of all plastic is for single-use items, tossed away almost as soon as they are acquired. The numbers surrounding our consumption of plastic are hard to fathom. And with production on an unceasing upward spiral, they're outdated as soon as one writes them down: Globally, a million plastic bottles are purchased every minute. *Five trillion* plastic bags are used each year. The average American goes through more than fifty kilograms of single-use plastic a year (only Australians match that level of wastefulness). The dollar figures involved are mind-boggling too. Globally, the market for take-out food containers, more than half of which are plastic, has surpassed $110 billion and is growing by 3.5 percent a year.

As a consequence of its pervasiveness in our daily lives, plastic is also omnipresent in the natural world. Astonishing examples are a regular feature of media reports. Microplastics—the tiny fragments into which most plastic eventually degrades—have made their way to the deepest oceanic trenches and the highest mountaintops. They are carried in clouds and riddle soil; entering our bodies with nearly every breath, bite, and sip; tainting our blood and penetrating our organs. One oft-quoted analysis predicted there would be more plastic than fish in the world's oceans by 2050. Another estimated the average human ingests a credit card's weight of plastic every week.

Roland Geyer has spent years trying to quantify plastic's ubiquity, and even he struggles to grasp the scale. I meet Geyer at the University of California, Santa Barbara, whose palm-tree-dotted campus juts into the Pacific Ocean beneath the towering Santa Ynez Mountains. His office opens onto an outdoor walkway overlooking a courtyard, and I can see the beach through a window on the opposite wall. A guitar is propped beside his desk, and an espresso machine sits nearby. Geyer has a close-cropped salt-and-pepper beard and wears tan buck-style shoes with bright blue laces and a lightweight hoodie with the sleeves pushed up. As a boy in Germany, he tells me, he'd watch garbage trucks collect rubbish on his street. "It seemed to me like I was the only one that wondered where it went and what happened to it," he says, and he only recently realized he's ended up building his career around those questions. He specializes in industrial ecology, a field devoted to mapping the flow of materials through the economy—tracing where our stuff comes from and where it goes. Early on, he studied steel and aluminum, but eventually noticed no one seemed to be doing "material flow analyses," as his approach is known, for plastic.

Soon, he connected with environmental engineer Jenna Jambeck, of the University of Georgia, and oceanographer Kara Lavender Law, of the Sea Education Association, and they began trying to determine how much plastic waste was entering oceans each year. As they pieced together that estimate, Geyer began wondering if it would be possible to answer some even bigger questions. How much plastic was there in the world altogether? What was it used for? And where was it ending up? Finding answers was "like putting a puzzle together," he tells me. The team gathered data from a variety of sources, using industry figures on production and environment agencies' estimates on waste, and then, where data was patchy, building models to extrapolate from what they knew. When the journal they'd submitted to returned their paper with peer reviewers' notes, one comment resonated for Geyer. It was a suggestion they expand their analysis to include plastic fibers—polyester and similar synthetics used in fabrics for clothing, carpets, and other textiles. "I thought, like, 'Oh, man,'" Geyer recalls, sighing and putting his forehead to his desk. "I knew I had to add it," but it meant months' more work.

The inclusion of both fibers and additives—chemicals that are mixed with a plastic's main ingredients to lend it specific properties—gave the group's estimate extraordinary comprehensiveness. And Geyer had realized something along the way. Because production data went back to 1950, and pre-1950 production was so small as to be essentially zero, the team would be able to put a number on all the plastic humanity had created. "I have never been able to do that in my research" on other materials, he says. It felt "really noteworthy and special," and to drive it home, he persuaded his colleagues to include the word *ever* in their paper's title. "Production, Use and Fate of All Plastics Ever Made" was published in 2017. The headline figure was mind-blowing: Industry had pumped out 8.3 billion metric tons of plastic. Of that, 6.3 billion metric tons had become waste, of which 9 percent was recycled, 12 percent was incinerated, and 79 percent ended up in landfills or the environment. If anyone should be able to grasp the scale of those figures, it should be him and his colleagues, Geyer says, but the numbers' vastness is such that "we're also struggling."

The data went deeper too. Forty-two percent of all non-fiber plastics went into packaging—much, if not most, of which was likely single-use. Nineteen percent was used in construction. Textiles accounted for more than 14 percent of total plastic production in 2015, the latest year covered by the study. The numbers provide a sharp view of petrochemical companies' ever-increasing output. The polyethylene Eric Fawcett and Reginald Gibson accidentally synthesized back in 1933 had been enough to "fill a hollow tooth," an Imperial Chemical Industries executive later said. In 1950, the industry made two million metric tons of plastic. Geyer pulls the 2023 figure from a spreadsheet for me: 513 million metric tons. Of course, the accumulated total keeps rising too. When he sees his paper's 8.3 billion metric tons "ever made" figure repeated, "I want to email them and say, 'No, it's actually 10 [billion] now because it's accelerating so relentlessly.'" Before I leave, Geyer clicks through a slide presentation to show me what the future might look like if production keeps climbing at current rates. By 2050, 1.1 billion metric tons of plastic would be made in a single year—and the cumulative total would be enough to cover the United States in a layer ankle-deep.

TO ROLAND GEYER, THAT PROSPECT FEELS HORRIFYING, BUT TO INDUSTRY it's simply a smart business plan. While concerned individuals worry about plastic's pervasiveness in our lives and its impact on the natural world, fossil fuel and petrochemical companies are ramping up to make more of it—not less—in the years to come. Other experts' projections corroborate the doubling Geyer noted would result from simply continuing along the current trajectory. One analysis even suggested plastic production could triple by 2050. This expansion, while stunning, represents not a departure from the past but a continuation of it. Today's plans are driven by the same forces that propelled the industry in John D. Rockefeller's day—unceasing hunger for profit, enabled by economic and political systems that allow such untrammeled greed to supersede all other considerations.

Now, though, there's new, even more pressing motivation. Oil and gas companies can see climate change's dangers will soon force the world—slowly and reluctantly, perhaps, but inexorably—to wean itself from their fuels, as we shift to electric vehicles and draw heat and power from the wind, sun, and other clean sources. Rather than putting their engineering and economic muscle into aiding that urgent transition, these corporations are betting big on plastic, hoping it will allow them to continue pumping oil and gas and keep profits flowing far into the future—regardless of the consequences beyond the pages of their earnings reports. "An escape hatch," as Carroll Muffett put it, from "necessarily declining demand" for fossil fuels.

No furtive whistleblowers are needed to tell this tale of corporate skullduggery. It's not a secret. On the contrary, it is the industry's openly declared plan, set out in black and white in companies' reports for investors, and their promises of ever-rising output. Often framed in dry corporate-speak, their strategy is nonetheless clear. Even ExxonMobil—the world's largest producer of single-use plastics, and among the most powerful and determined opponents of a global shift away from fossil fuels—acknowledges electric vehicles' widespread adoption will likely push cars' need for oil significantly lower by 2050. In a detailed market analysis, the company said it expected chemical production to increase by 80 percent over the period, and predicted that, together with demand from shipping, trucking, and aviation, plastic and other petrochemicals would allow industry to pump even more oil in 2050 than it does today. "Chemicals and commercial transportation account for almost all of the [oil] demand growth," the company projects. With no acknowledgment of the vast amounts of plastic put to wasteful use, ExxonMobil justifies its plans with platitudes about the material's importance, touting its intention to sell ever more of it in developing nations whose current consumption is just a fraction of wealthy countries'. Demand will continue growing, the company says, "as more of the world's people advance to the middle class and gain access to products essential for modern living."

The International Energy Agency echoes that view, projecting that while use of oil for fuel will soon begin falling, "the steady pace of growth for petrochemicals will not relent." It noted an "explosive increase" in production capacity in the early 2020s and predicted even more going forward. (It's worth noting here that plastic and its ingredients account, by one estimate, for 80 percent of petrochemicals.)

That steep upward climb is not inevitable, though. There are obstacles on the path industry envisions, and they offer opportunities to those hoping to change the trajectory. Like oil, plastic making is a cyclical, boom-and-bust business, riding the roller coaster of economic peaks and troughs. A frenzy of plant building over the past decade has created a supply glut that's prompting some companies to pull back temporarily from plans for more. Shell in 2021 envisioned transforming its oil refineries into "Chemicals and Energy Parks" as it leaned further into petrochemicals. The company, which expanded its plastic-making capacity by nearly 150 percent in the first half of the 2020s, touted a new polyethylene complex in western Pennsylvania and expansions of already-vast plants in Louisiana and Guangdong province, China, together increasing by millions of tons the volume of plastic it makes annually. As is common for the industry, Shell framed it all as a response to demand rather than its own choice. "We must give our customers the products and services they want and need," the then-CEO said. By 2025, the company was hinting it might back away from chemicals, with a new boss worrying they were "not delivering adequate returns."

But if one company sells its plants, another will buy them. And it's clear the industry as a whole sees plastic production going only in one direction—up. With the International Energy Agency predicting petrochemicals will be the largest single driver of oil-demand growth this decade, plastic is no longer just a handy way to get profit from what would otherwise go to waste. It is fast becoming one of the main economic justifications for continued drilling.

Indeed, that helps explain why plastic's relentless growth threatens to push us further along the path toward spiraling, out-of-control climate change. By giving companies another way to monetize oil and gas, it adds to their incentive to keep pumping more, much of which will still get burned, even if a portion is turned into plastic. It's clearer with every passing—hotter—year that anything other than keeping those fuels in the ground adds to the grave danger we already face. What's more, the process of turning oil and gas by-products into plastic delivers its own hefty wallop of carbon dioxide. Shell's Pennsylvanian polyethylene plant, for example, is allowed to pump out planet-warming gases equivalent to those from 480,000 cars. And a great deal of plastic ends up incinerated, bringing an additional climate hit. There is also worrying evidence that plastic scattered in the environment may emit greenhouse gases as it degrades—a vast, potentially uncontrollable source of emissions—and even more consequentially, that it interferes with the oceans' crucial role in slowing global warming by soaking up carbon dioxide.

On top of their climate dangers, of course, industry's plans, if realized, will also mean a future stuffed with ever more plastic packaging, throwaway utensils, fast fashion, and other junk that's tossed away almost as soon as it's used. Experts warn of a lock-in effect, in which new plants, once built, will keep churning out plastic for decades. The up-front investment in construction is hefty, while operating costs are relatively low, so companies have a strong incentive to run plants at maximum capacity for as long as they can—making, and selling, as much plastic as possible.

For the industry that taught Big Tobacco to obfuscate, these plastic dreams are in many ways just another form of denial—the latest iteration of the thinking that's guided them for decades. It's not only that the big bet on plastic is rooted in efforts to downplay and discredit climate science. The plans are also key to Big Oil's strategy for prolonging the life of its destructive business model. After all his years rooting through fossil fuels' history, Carroll Muffett, for one, is unsurprised to see industry bringing the same strategies it's long used to perpetuate oil and gas to bear on behalf of plastic. Because, as he reminds me, "it's literally the same companies."

CHAPTER 2

"An Excellent Salesman"

1958—Lego patents its interlocking system for toy blocks. Six million metric tons of plastic produced.

Adolphus Green never meant to run a biscuit company. He was a Chicago lawyer when, in 1890, a group of cracker and cookie makers came to him for help merging their companies. Green shepherded through the deals that turned forty bakeries in thirteen states into the American Biscuit and Manufacturing Company, then went back to his legal work. It was an era of corporate consolidation, with new behemoths born in sectors across the economy, and the same thing happened in baked goods. By 1898, having swallowed up many of its competitors and renamed itself National Biscuit Company, the firm Green had helped create controlled half the country's cracker and cookie business.

When the board of directors put him in charge, he decided the company needed a marquee product, and he settled on a soda cracker. At the time, nearly all the country's bakeries shipped their biscuits locally, in large barrels from which grocers and general store owners doled out portions to customers. The same system was used for goods like sugar, flour, and molasses. Green had a new idea. Individual packages would connect the manufacturer more directly to shoppers, and lengthen crackers' shelf life, so they could be produced centrally and shipped long distances by rail. His law partner, Frank Peters, spent months playing with designs, eventually layering waxed paper and cardboard into a cleverly folded carton. The new brand, Uneeda Biscuit, debuted in 1899.

The packaging not only kept the biscuits crisp and clean—neither was a given in barrels—but it also enabled the company to link its product to a national advertising campaign, the country's biggest to date. "With the forefinger of your right hand, point to the shelf and say to the grocer 'I want these,'" one ad instructed. The campaign's central figure, a boy in a yellow rain slicker and hat, became one of America's most recognizable images. He earned his keep. Within a year of Uneeda's debut, the National Biscuit Company—its name was later shortened to Nabisco—was selling ten million packages a month, twenty times the *annual* sales of all other packaged crackers combined.

It was a watershed moment, the beginning of the end of the "cracker barrel" era of bulk food sales by small local merchants, and the first glimmer of a system that would come to be dominated by national brands with huge reach, and eventually big supermarket chains too. "The new package changed the way people shop," letting them "touch and examine the box" and "form a relationship with the package that extended from advertising to the shop to their home," one packaging expert later wrote. Other companies quickly followed. Soon Quaker Oats was selling cereal in cartons, with its own big ad campaign, and by 1925 nearly seven hundred thousand tons of boxes were used to sell everything from candy and ice cream to butter and soap.

From the start, consumer packaging combined meaningful practical advantages—it kept food fresh and sanitary, enabling contents of uniform quality to be manufactured in large quantities and stored for extended periods—with big marketing benefits. Retailers could stack packaged goods in eye-catching piles, confident customers would recognize brands they'd encountered in magazine, radio, and, later, television ads. Clever logos and colorful designs would grab shoppers' attention, and labels touted products' wonders. Companies had learned that packaging could be highly effective at "motivating sales and repeat sales," one packaging design executive later explained, making it "a major source of the planned growth of product lines and profitability."

IT WAS TRUE FOR DRINKS TOO. BUT UNLIKE THROWAWAY CARTONS, THE glass bottles in which soft drink and beer makers sold their products could be used again and again. Customers paid a deposit with every purchase, then reclaimed it when returning the bottles for cleaning and refill. Coca-Cola helped take that system national in the early decades of the twentieth century, relying on a network of local bottlers, franchises that mixed its syrups into beverages they distributed across designated territories. That decentralized system of reuse suited everyone. Head office was happy to offload the logistics of selling in remote, rural parts of the country. Bottlers liked returnables because glass was expensive, and they "simply could not afford to have their valuable investments thrown in the trash," as business historian Bartow Elmore wrote in *Citizen Coke: The Making of Coca-Cola Capitalism*. Deposits of a couple pennies on top of a drink's three-cent price reliably got most containers back. In the late 1940s, the soda bottle return rate was 96 percent; some were refilled fifty times.

By then, the beer industry had already embraced what were known at the time as "one-way containers." Prohibition had decimated small brewers in the 1920s, and the big fish who survived into the '30s were eyeing markets vacated by local businesses' demise. Returnable bottles were an obstacle. Because a truck could collect only so many empties before turning around, reuse created "a natural limitation on the market area served by a bottling plant," Peter Chokola, president of his family's

Wilkes-Barre, Pennsylvania, bottling firm, later explained to a Senate committee. So behemoths like Anheuser-Busch and Miller jumped on steel cans, which became available around 1935, despite an awkward design that required drinkers to use a special key to poke a hole in the top. "Consumers were largely satisfied with returnable bottles" and had not asked for the change, a big can maker wrote years later. But throwaway cans—and, more fundamentally, the idea of disposability—allowed corporate giants to ship huge volumes of beer across the country. Just as individual boxes had broken the link tying biscuit sales to local hubs, one-way containers paved the way for beverages to become a national—and soon, international—business dominated by a handful of companies.

It took a little longer for soda, whose acidity ate away at cans, and whose carbonation created pressure that burst them. By the 1950s, designs had improved, but big companies like Coke were wary of a change that would force their bottlers to buy pricey new equipment. Walter Mack wasn't afraid of innovation. He'd built Pepsi-Cola from an also-ran into Coke's top competitor, then left to run a small drink maker called C&C, where he started selling canned soda in 1953. The company pitched retailers with cartoon depictions of the hassles throwaways would free them from—shopworkers sorting empties and sweeping broken glass, insects flocking to spilled soda.

The cans made a splash. From across the Atlantic, Britain's *Daily Mirror* proclaimed "a wholly new way of drinking that makes old-style bottled beverages as obsolete as yesterday's horsecars." At Coke's Atlanta headquarters, executives now saw one-way containers could end their dependence on local bottlers, enabling them to consolidate distribution and gobble up the middlemen's profits. The company began experimenting with cans in 1955, and their use quickly grew, with aluminum replacing steel after 1967; it continued using glass bottles too, but now they were meant to be discarded, not returned. For small bottlers, it was devastating. Disposables "provided the medium through which monopolization of the soft drink industry could be achieved," Chokola, the Pennsylvania distributor, said in urging senators to ban them. Indeed, the 5,000 soda bottlers that had dotted the country in 1947 dwindled to 1,600 by 1970, with more closing every year. Consolidation was even more dramatic for breweries, whose numbers dropped by three-quarters in just sixteen years.

For Coke and the few big bottlers still standing, profits only grew. Freed from bottle collection and cleaning, they saved on labor. With shipments going just one way, trucking got cheaper too. Grocery stores also appreciated not having to deal with empties. "They're a pain in the neck," one shop owner said. Of course, the costs didn't disappear. Coke—and the competitors that followed its lead—had simply pushed them onto someone else. Municipal sanitation departments, Chokola explained, were now saddled with the "recovery burden of sixty-five thousand truckloads" of cans and bottles a day. "The consumer is the one who pays the taxes to clean up the mess." Indeed, from the companies' perspective, that was the beauty of the new system.

Disposables cost the public more directly too. Coke was 30 to 40 percent more expensive in throwaway cans and bottles than in reusables, the head of its US business said in 1972: "The difference lies essentially in the different costs of packaging." While a returnable bottle's cost was spread over many purchases, a throwaway's price "is absorbed in one use," said the executive, J. Lucian Smith. In Richmond, Virginia, for example, a six-pack of canned Coke cost eighty-three cents; the same six-pack in returnable bottles was fifty-nine cents. Not only did reusables "offer the best value to the consumer," Smith said, but they also "provide the most ecologically sound method of distributing soft drinks" and "significantly alleviate" communities' waste problems. Consumers noticed the price difference and seemed to prefer returnables, he acknowledged—although his company's ads were claiming the public had demanded disposables. In the end, it wasn't up to consumers. As

Chokola put it, the beverage companies "have removed" the cheaper option "from the realm of consumer choice." Returnable bottles were vanishing.

All the while, in a lab at the DuPont chemical corporation, Nathaniel Wyeth—brother to the painter Andrew—was tinkering with plastics that might be strong enough to hold soft drinks, whose carbonation meant they required particularly strong bottles. Pepsi was first to embrace the material he finally settled on, called polyethylene terephthalate, or PET. The number two soda maker brought PET bottles to market in 1976, while Coke put its muscle into a different plastic, acrylonitrile, made by Monsanto. Coke dubbed its bottle the "Easy-Goer," to tout its light weight. But troubling questions surrounded it from the start, and while Coke and Monsanto insisted their bottles were safe, the Food and Drug Administration banned them in 1977, saying they could leach a chemical that caused cancer in rats. Coke switched to PET, whose low cost and all-but-endless availability would serve as rocket fuel for disposability. But the path to the beverage giant's current annual global sales of a nearly unfathomable 137 billion plastic bottles had been laid long before. And while the throwaway ethos had been invented and honed in the United States, it would soon be exported around the world.

IT WASN'T JUST FOOD AND DRINKS, OF COURSE. THROWAWAY PACKAGING could be used for just about anything. And it not only helped sell whatever it held; it also became a source of profit itself. Chemical giant DuPont boasted in 1956 that a big dry cleaner's adoption of plastic wrapping had turned into "one of the hottest promotions" the firm had run. Within two years, the bags' makers had sold a billion of them, raking in $20 million. The Standard Packaging company tripled its sales between 1955 and 1958, entirely because "everything we make is thrown away," its director explained. Polystyrene foam—commonly known as Styrofoam, the name Dow Chemical trademarked for it—appeared in the '50s too, and was soon being used for coffee cups, egg cartons, meat trays, take-out containers, and more.

Lloyd Stouffer was one of the first to understand the profits disposable packaging—and plastic's role in it—promised. Stouffer was editor of *Modern Packaging* magazine, *Modern Plastics'* sister publication, and in 1956 he laid out his vision in a speech to the Society of the Plastics Industry. "The future of plastics is in the trash can," he told the assembled executives. Ditching reuse to "concentrate on *single* use," he explained, would transform "a one-shot market for a few thousand units" into "an everyday, recurring market measured by the *billions* of units." To realize that potential, the industry would have to, as one writer later put it, "teach customers how to waste," weaning a generation that had endured years of deprivation off the ethos of reuse and frugality.

That process had already begun. Three years after the first squeeze bottle was produced, in 1947, its manufacturer was making one hundred thousand a day—containers for products like hand cream, suntan lotion, and shampoo. By 1963, when Stouffer returned to address the plastic producers again, he was thrilled to report further progress. "You are filling the trash cans, the rubbish dumps and the incinerators with literally billions of plastics bottles, plastics jugs, plastics tubes, blisters and skin packs, plastics bags and films and sheet packages," he crowed. "The happy day has arrived when nobody any longer considers the plastics package too good to throw away."

Across the Atlantic, Stouffer reported, more than a third of cooking oil was being sold in throwaway bottles, "a revolutionary change for Europeans who are accustomed to getting this product in heavy, returnable glass bottles." At home, high-density polyethylene's widespread use for detergents and cleaning supplies previously sold in glass was "so well known that I don't think it's necessary for me to comment." Amber-colored plastic was making inroads with pharmaceuticals, an industry that had been using three billion glass bottles annually. Budweiser and other brewers were switching from paper to plastic six-pack carriers, saving a penny on every one. Soon, plastic would displace cork as the lining for metal bottle caps ("small in size but big in volume") and provide lids for cans and tubs ("the potential on ground coffee alone is one billion" a year, Stouffer said). The new material was taking over so fast, predictions were unreliable. One expert "suggests that your market in packaging will double in the next two years," he said. "Another, highly cautious, sees it as tripling by 1970." In the years to come, it would grow far beyond even those lofty hopes.

BY THE LATE 1960S, THOUGH, DISPOSABILITY'S DOWNSIDE HAD BECOME hard to ignore. "Major U.S. Cities Face Emergency in Trash Disposal," a typical headline warned, over an article citing a federal estimate that upgrading waste management systems to keep pace with the growing volumes would cost $2 million a day for five years. And producing all that packaging was consuming vast amounts of natural resources—bauxite for aluminum cans, trees for paper and cardboard, and petroleum derivatives for plastic. It sucked up lots of energy too. Making Americans' bottles and cans alone used the equivalent of ninety-two thousand barrels of oil a day—more, by one estimate, than the combined consumption of fifteen African, Asian, and Central American nations. So in 1969, two months after Neil Armstrong planted a nylon flag on the moon, industry and the federal government sponsored the first national conference on packaging waste. The San Francisco gathering drew representatives from dozens of companies that made and used packaging—can and bottle producers; paper corporations; big oil and chemical firms like Dow, DuPont, and Mobil; food and drink sellers including Coca-Cola and Pillsbury; and consumer products companies such as Johnson & Johnson and Procter & Gamble.

"AN EXCELLENT SALESMAN"

Despite the ostensible focus, most speakers extolled disposable packaging's benefits, for both consumers and their own bottom lines. "Let us remember the pre-packaging kitchen, the American woman giving her youth, health and strength to feed her menfolk," packaging design executive Eric Outwater exhorted. The containers that so eased her lot—by enabling a cornucopia of new processed foods—had exploded into a $31 billion business, from just $2 billion three decades earlier, he pointed out. In beverages alone, "each deposit-type bottle displaced from the market means the sale of twenty one-way containers," he said. "We are now in an era of open-ended opportunity called convenience packaging."

Another speaker, a consultant analyzing waste, reminded listeners why packaging's presence kept growing: "because the package is such an excellent salesman." For example, he explained, "it is easier to sell a small item, say a fountain pen, when it is sealed in plastic against a large, colorful poster." Indeed, Outwater added, "the development of a packaging concept has been known to open up entirely new product areas." Soon, he predicted, Americans would be munching cereal out of "eat-from, throw-away" bowls and buying single servings of condiments, soups, and entire meals. The possibilities seemed unlimited, and a Dow Chemical executive took Outwater's vision even further. "Mass feeding systems in industrial cafeterias, and institutions such as universities, hospitals, airlines, restaurants, etc., are adopting completely disposable service-ware of plastics and plastic-coated materials," he said. It was a prescient description of things to come.

To another conference speaker, that future didn't look so sunny. Leonard Stefanelli stood out at the event, for both his background and his blunt manner. He'd been just seven when, at the tail end of the Great Depression, he started ringing San Francisco doorbells to collect payments—fifty-five cents a month per household—for his uncle's trash business. By 1953, Stefanelli, then a skinny nineteen-year-old who'd barely made it through high school, was a garbage collector himself, hauling fifty-kilogram cans of stinking trash into an open-backed truck. "I actually loved working on the truck," he wrote later. Every bar on the route would give the crew beer, or coffee spiked with booze. "There were not many of these," he recalled, "but enough to satisfy a thirst." Soon he was taking night classes at college, joking to friends that he went to the University of San Francisco twice a day, "once to get an education and the second time to pick up the garbage."

Stefanelli had been on the job for twelve years when workers took control of Sunset Scavenger's board and made him president of the company, one of just two handling the city's trash. The august crowd at the packaging conference didn't intimidate him. He described how, when he'd first started out, other than engines replacing horses, garbage trucks' setup hadn't changed much in decades. One man stayed in the back, pulling salvageable material—rags, cardboard, metal, glass bottles—from an ever-growing heap of trash. In 1954, the company had made half a million dollars selling such items. That turned out to be the peak. As he rolled up and down San Francisco's steep streets, Stefanelli told his audience, he'd seen "radical changes in the physical composition" of the city's trash. "This was the beginning of the 'no deposit, no return' era of packaging materials." Sunset Scavenger had to dump tens of thousands of refillable champagne and beer bottles when producers switched to throwaways—still glass in those days—which themselves didn't fetch enough to be worth selling. Old clothing had once gotten fifty cents a kilogram, but when natural fibers were replaced with synthetics that repelled moisture instead of absorbing it, there was "no demand for these type of rags," so they went to landfill too.

It wasn't just the makeup of the garbage that had changed. Its volume had jumped by 50 percent in a decade. Instead of one bin, by the late '60s, "we have two and sometimes three cans of refuse" from the same-size household, an increase that "can be attributed only to the packaging industry." Plastic, Stefanelli said, was "probably the single largest contributor."

Like collectors across the country, Sunset Scavenger had needed new compacting trucks to compress the huge amounts of trash, Stefanelli explained. That made it impossible to retrieve anything salvageable, even if there had been a market left for it. "To summarize, all I can tell you is that the packaging industry has created one hell of a situation," he said. While the problem was global, he believed America's role was unique. Having "charged forward" with inventing "beautiful packaging concepts" to attract consumers, it had created a new problem: "what to do with these 'beautiful products' after they have been used only once."

The former garbageman wasn't the only one at the 1969 conference to warn about disposable packaging's dangers. William Gunn, chief of the packaging design firm Stuart and Gunn, for instance, worried the waste problem could become an "apocalyptic monster," and said he did not "wish to see my children or grandchildren choke in the mess they have inherited from us."

One way to grasp what disposability meant for waste, packaging consultant Arsen Darnay said, was to compare the number of bottles made to the number filled. In 1958, he said, twelve billion had been manufactured, but bottles were filled fifty-three billion times, meaning, on average, each was used more than four times that year. In 1966, the typical container was filled twice. "By 1976, the ratio will be roughly one to one, and returnable beverage containers will have almost disappeared," Darnay predicted. More and more of the resulting waste would be plastic. Bottles had consumed 152,000 tons of the stuff in 1966, a figure Darnay projected would more than quintuple in a decade. And drinks, of course, were just one sector. "If the US consumer could curb his appetites, and if package consumption per capita" stayed flat, the country would generate millions fewer tons of waste annually, saving vast sums for localities tasked with managing it, he said. "But this is unlikely to happen."

The astonishing gall of that assertion packs a wallop across more than half a century. It was Darnay who'd described how swaddling a fountain pen in packaging helped sell it, and he'd just explained how reusable options were vanishing. Yet there he was, in front of a roomful of men making handsome livings from packaging, citing individuals' "appetites" as the culprit. It wasn't the last time industry would blame consumers for the changes it pushed onto them. And Darnay, whose consulting clients had included Coca-Cola, would soon become a top federal waste official—a telling illustration of the industry's influence over what government would be willing to do to rein in disposables.

For the most part, it fell to conference speakers from outside industry to engage seriously with solutions—especially those that could dent companies' profits. New York's deputy sanitation commissioner Hugh Marius suggested that a fee imposed on producers might rectify the dynamic that made packaging so cheap: Its price included neither the cost of dealing with the waste it quickly became nor its impact on the natural world. Such a levy—collected by the federal government and distributed to municipalities—would not only create "economic incentives to reduce the most troublesome kinds of packaging" but also fund trash collection and subsidize recycling, Marius explained. "I hope that you agree with me that serious consideration should be given" to such a charge, he told the industry audience.

They most certainly did not. Decades later, that idea—known today as "extended producer responsibility"—remains at the center of bitter fights over who should bear the true costs of disposability. For the most part, the other speakers' exhortations were ignored too. The companies making so much money from throwaway packaging were not prepared to give it up.

One warning sounded repeatedly at the 1969 conference was that if industry failed to act, government might step in with costly ideas like Marius's. To head off such unwanted interference, Republican Congressman Paul McCluskey told the gathered executives, they had to do something about the waste they were creating—"not dispose of it, but reduce the actual amount." Rising public concern about the environment "should not be underestimated," and could, before long, prompt "sweeping changes in our laws." Soon, Congress would probably consider requiring a five-cent deposit on bottles, he said. While McCluskey acknowledged politicians' consideration of tougher action was driven by public opinion, he nonetheless feared that as "the pendulum swings to the protection of the environment," there was a danger "the swing can be too far."

He needn't have worried. In the years to come, companies would go to extraordinary lengths to protect the lucrative model they'd created, even as the mountains of trash it churned out grew ever larger.

CONGRESSMAN MCCLUSKEY WAS RIGHT ABOUT THE COUNTRY'S MOOD. THE packaging waste conference coincided with a growing awareness of consumerism's dark side that crystallized with the first Earth Day, in April 1970. In Atlanta, where an underground newspaper urged readers to "bring the trash home to the people who make it," 1,500 protesters showed up at Coke's head office with bags—and a pickup truck—full of empty cans and bottles. Even President Richard Nixon got on the bandwagon, decrying the shift toward disposable packaging. "We often discard today what a generation ago we saved," he said, noting that "pouring more and more public money into collection and disposal of whatever happens to be privately produced and discarded" amounted "to a public subsidy of waste."

Plastic executives feared the upswell of antagonism might "really end the industry," *Modern Plastics'* editor later recalled. But they were ready to fight. In 1954, after Vermont prohibited disposable beer cans—farmers blamed them for damaging equipment and injuring cows when they were tossed into fields—nearly two dozen beverage, cigarette, candy, and packaging companies had formed an alliance they called Keep America Beautiful. In billboards and print ads, the group framed the country's waste issues as a matter of littering, the result not of excessive production, but rather the bad habits of irresponsible individuals. Keep America Beautiful printed handbooks to teach schoolchildren "good outdoor manners," distributed "how-to-do-it" litter-reduction kits, and instructed drivers to keep trash bags in their cars instead of tossing empties out windows. Under pressure from brewers and can makers, Vermont let its ban expire in 1957. But the organization whose creation it had prompted was just getting started.

No matter the medium, Keep America Beautiful's message was clear. The problem wasn't the overwhelming amounts of waste the shift to throwaway packaging was generating, but merely that a fraction of it was ending up in the wrong place. In other words, the mess wasn't the fault of companies reaping huge profits from disposability, but of Americans themselves. And responsibility for fixing it was on them too. "Keeping America clean and beautiful is your job," one ad lectured. Another accused readers of hiding behind the "alibi" that "average people don't pollute. It's the corporations," and—with impressive nerve—exhorted them to "stop shifting the blame." That very American message of individual responsibility—and its failure to acknowledge larger economic and political forces—was not unique to the issue of waste. Indeed, one packaging executive's assertion that "packages don't litter, people do," carried eerie echoes of gun rights proponents' claim that "guns don't kill people, people kill people."

Companies hammered home the idea in their own ads. One Coke campaign, "Bend a Little," featured an attractive woman bending over to pick up litter, "to remind people that cleaning up America called for a little extra effort from all of us." Another favorite theme was that consumers, not drink makers, had driven the shift toward disposability. The public "demands a choice of containers in many products, including soft drinks," Coke explained in one ad, as if it were a helpless actor in the change it had pioneered. "We have to go along." Not only were individuals responsible for littering; the entire system of throwaway containers was their fault too.

By 1967, with public concerns about waste nonetheless mounting, Keep America Beautiful decided to take things up a notch. "Our 'soft sell,'" executives at one planning meeting concluded, should be toughened to slam litterers as "slobs." Before long, a new TV commercial showed pigs trotting down a rubbish-strewn street, then nosing through garbage on a beach, as a sanctimonious voice-over said, "Spreading and living in litter is for . . . Well, certainly not for people."

With Earth Day 1970 bringing the threat to yet another level, companies had to up their game again. Keep America Beautiful's ad team got to work on a spot that would become part of TV history. In "The Crying Indian," which debuted in 1971, a man with a craggy face, tasseled buckskin jacket, and long braids paddles a canoe to a riverbank covered with trash, then walks to the side of a highway. When a bagful of garbage thrown from a passing car lands at his feet, he looks into the camera with a tear on his cheek. "Some people have a deep, abiding respect for the natural beauty that was once this country," a narrator intones over dramatic music. "And some people don't. People start pollution. People can stop it."

Producers realized its power immediately, "Iron Eyes" Cody, the actor who portrayed the Crying Indian—he was actually Italian American—later wrote. "People who had been working on the project were moved to tears just reviewing the edited version," he recalled. It was "an iconic moment" that "was like nothing else on television," one media scholar concurred. Audiences agreed. The ad achieved one of the highest viewer-recognition rates in history, airing so often that within a few years, broadcasters were ordering replacement films because they'd "literally worn out the originals from the constant showings." It was showered with awards and, even more telling of its place in popular culture, was later spoofed on *The Simpsons* and referenced on *Friends*. Cody's face appeared on billboards and in print too, and new iterations of the commercial ran throughout the '70s. In an extraordinary twist for a character cooked up by polluting industries, "the Crying Indian became for millions of Americans the quintessential symbol of environmentalism," author Finis Dunaway wrote in his book *Seeing Green*.

Much of the campaign's power came from viewers' perception of Keep America Beautiful as an impartial civic organization. Its corporate sponsors—they included Coke, Pepsi, Philip Morris tobacco, and the American Can Company—didn't trumpet their involvement, and few knew of it. The ad's genius, Dunaway observed, was that it "promoted an ideology without seeming ideological; it sought to counter the claims of a political movement without itself seeming political." And with "the guilt-inducing tear," it managed to "propagandize without seeming propagandistic." The ad deftly weaponized stereotypes of Native Americans' connection to the land—and their sorrow—to intensify its punch and lend a counterculture feel. Industry had "co-opted the icon of resistance and made him support the interests of the very consumer culture he appeared to protest," journalist Ginger Strand noted. "There he stood, stoic and sad, a rebuke to individuals rather than a rejection of the ideology of waste." Just as important, the spot neatly shrank disposability's harms down to litter alone. If consumers would simply dispose of throwaway containers properly, it suggested, everything would be fine.

"The Crying Indian" had helped muddy public understanding of what was causing the waste deluge, what might fix it—and even what the problem was. That intentional sowing of confusion made it a defining prototype for the burgeoning genre of greenwash, the deceptive set of marketing tactics aimed at imbuing products with an aura of environmental friendliness. And it perfectly prefigured the concept of the "carbon footprint," an industry-propagated framing that focuses us on our own contribution to climate change, and consequent feelings of guilt, while ignoring the ways corporate and political leaders' decisions have constrained our choices.

The goal was clear. If a group like Keep America Beautiful could persuade Americans to swallow its message, author Heather Rogers explained in *Gone Tomorrow: The Hidden Life of Garbage*, it would mean "laws will not be enacted, government won't intervene, and production can continue on industry's terms." She pointed out that "the way Americans waste today is not just a normal result of organic human development." Vast and tremendously profitable corporations had set out to create the throwaway world we live in. Now they were blaming us for its consequences.

Like disposability itself, the personal-responsibility framing would soon be exported, as industry plowed money into anti-litter groups such as Keep Australia Beautiful, Keep Britain Tidy (whose first chairman ran an oil company), Gestes Propres (Clean Gestures) in France, and Paisaje Limpio (Clean Landscape) in Spain. The UK got its own heartstring-tugging TV spot, in which a stirring rendition of Britons' much-loved hymn "Jerusalem"—whose lyrics imagine Christ visiting their "green and pleasant land"—plays over images of a road heaped with rubbish. "Britain is a beautiful country, not a litter bin," a narrator scolds. A British Rail poster exhorting riders not to leave trash at stations featured the Swedish pop stars ABBA in yellow "Keep Britain Tidy" T-shirts—sparkly bodysuits presumably left at home—holding brooms and cleaning gloves.

The early 1970s upswell of environmental worry was the first of what would become regular moments of crisis for the industries making and using throwaway packaging and goods. With Keep America Beautiful and "The Crying Indian," companies had created an effective template for defusing and deflecting anger about waste—embracing and co-opting environmental concerns, then directing blame back onto the public. They would turn to it again and again in the years to come.

WHILE INDUSTRY DRAPED ITSELF IN GREEN LANGUAGE, IT WAS FIGHTING furiously against measures that could actually solve the problem it purported to be so concerned about. In 1971, Yonkers, New York, like cities across America, was struggling to manage its growing tide of trash. Not only was garbage strewn around streets and parks (Keep America Beautiful wasn't wrong about litter being a problem; it just wasn't the only problem), but landfill space was getting more expensive. A student committee appointed by Mayor Alfred DelBello suggested a solution that had been bandied about repeatedly in state legislatures across the country—and in Congress too—but only rarely enacted. The panel's proposal would require beverage companies to include a five-cent deposit in the cost of bottled and canned drinks; customers could reclaim it when they returned containers for refill or recycling. Projections estimated it would cut the city's trash hauling costs by a million dollars a year, and save Yonkers shoppers $2 million annually on drinks. DelBello, a Democrat, cosponsored the bill with the city council's top Republican, and it looked to have a good shot at passage.

Then industry showed up. Suddenly, Yonkers, a city of two hundred thousand just north of the Bronx, found itself on the receiving end of intensive lobbying by soft drink, beer, and packaging interests. At a marathon city council hearing, Sidney Mudd, president of a big 7-Up bottling company, said the soda was no longer available in returnable bottles in the New York area, and vowed he'd stop selling in Yonkers before spending what he estimated would be $21 million to retool his plants for returnables. If the plants had to close, five hundred jobs would go with them, he said. Representatives of Coke, Pepsi, the US Brewers Association, a state grocers' alliance, and major can and bottle makers echoed his message. "The young people who spoke in favor of the bill were well-prepared," Councilman Peter Mancusi told *The New York Times*. "But they couldn't put on a show like this. It was like a Broadway play." A bill that had seemed like "a noncontroversial, beneficial, ecology-minded thing" had become explosive, the mayor said. The politicians were spooked by industry's predictions of a huge economic hit. In any case, one councilman said, "I learned that companies are understanding and facing the waste problem themselves." And "who can solve problems better than private industry?" Some lawmakers swung behind the plan again later, after other bottlers told the students shifting to returnables wouldn't kill jobs after all. But the bill never came back to the full council for a vote.

Similar stories played out across America. In Washington state, a container-deposit ballot measure polled at 80 percent weeks before the vote, but lost 51–49 after industry poured $2 million into ads opposing it. Supporters, by one account, spent only $6,000. By 1972, more than 350 state and local bills mandating deposits or otherwise restricting throwaway beverage containers had been introduced, but only a couple of dozen had passed, in places from Davis, California, to Madison, Wisconsin, and Edgewater, New Jersey. Even many of those were ultimately repealed under pressure or reversed by courts following industry lawsuits. Among the first to be struck down was a two-cent tax New York City slapped on plastic containers in 1971.

The deposit laws, known as "bottle bills," were anathema to industry because they pushed the costs of dealing with used packaging—which disposability had so successfully off-loaded to municipal governments—back onto companies. Thrilled to have freed themselves from the hassle and expense of handling empties, they were determined not to go back. One Keep America Beautiful executive slammed bottle-bill advocates as "communists" and urged the group to resist such laws.

In 1971, Oregon became the first state to enact a bottle bill, and Vermont soon followed. Both measures survived court challenges, and a year after the Oregon law took effect, an analysis found roadside litter had decreased by 35 percent, nearly four hundred million fewer cans and bottles were used even as drink sales kept growing, and the energy savings were enough to heat fifty thousand homes. What's more, the law was extremely popular—91 percent of Oregonians approved.

But passing bottle bills proved tremendously difficult. The same day as the testy 1971 Yonkers hearing, state lawmakers across the Hudson River in New Jersey were considering a similar law. Americans, its Republican sponsor said, could save $1.5 billion a year on drink purchases if all beer and soda containers carried deposits. But his bill soon ran into trouble, with even some cosponsors rescinding support after bottle makers said it threatened thirty thousand jobs in the state's glass industry. The claim was false. Given its reduced labor requirements, a state analyst explained, it was disposability—not a deposit-and-return system—that would be the job-killer. It didn't matter. The bill was dead.

Industry's biggest worry was a national bottle bill. Congress took up the idea at least five times in the 1970s alone. One analysis found such a law would save the equivalent of nearly twenty million liters of gasoline a day. Another estimated it would reduce drink prices by nearly a third, and increase—not decrease—employment. Nearly three-quarters of Americans approved of the idea, and in 1974, the Nixon administration backed it. In response, Pepsi's president fired off a letter to the head of the Environmental Protection Agency: "Your position defies and denies the free will of the people expressed by their free choice of containers," he sniped. In 1979, as the House considered yet another bill, an executive from one of Coke's largest bottlers said the Oregon law had more than doubled his company's fuel costs. "Left unsaid was the fact that municipal agencies had been paying the fuel costs of picking up the company's waste," the Coke historian Elmore noted.

"AN EXCELLENT SALESMAN"

Bill after bill would be proposed in the years that followed, but a national container deposit law never came to be. By the mid-1980s, ten states had enacted bottle bills. Since then, only Hawaii has done so, and with Delaware repealing its law, the total today still stands at ten. In Washington state, where the 1970 container deposit referendum lost narrowly, a 2024 headline carried more than a bit of déjà vu: "Bottle Deposit Proposal Fizzles Out in Legislature."

Where bottle bills did make it into law, industry often sought to undermine their implementation. A Vermont resident complained labels were intentionally confusing, writing to a magazine that "many bottles have NO DEPOSIT NO RETURN *cast* into their sides, but have the *deposit* notice on the bottom." New York, after passing its bottle bill in 1982, required the adoption of reverse vending machines, where consumers could return bottles and cans and claim their deposits. A Norwegian company tried to introduce them, but Coke argued the scanning technology would be unable to distinguish among brands, making it impossible for companies to know how much they owed.

Container deposit laws are common in Europe. For Scandinavians, "returning your empty bottles to the store is a perfectly normal, everyday activity" for everyone from hungover students to kids seeking pocket money and families heading to the grocery store, environmental historian Finn Arne Jørgensen wrote. The same is true in Germany, which—with a twenty-five euro cent deposit on plastic bottles—gets 98 percent back.

But around the world, industry continues to fight such laws. In Australia, a New South Wales lawmaker said a lobbyist for food, drink, and packaging companies had threatened a $4 million ad campaign against the Liberal and National parties ahead of 2011 elections if they backed a container deposit proposal. The politician, Catherine Cusack, called it "a nasty experience" that was "one of my worst moments in the job." The Liberals dropped their bottle bill, although New South Wales, Australia's most populous state, passed one a few years later. In Edinburgh, Coke lobbied against a mandatory bottle deposit plan, arguing that "consumers don't want" such a law, even as a poll found three-quarters of Scots supported the idea. The measure passed in 2019, but its start date was repeatedly postponed, in part because of disputes with Britain's Conservative national government, which said its details needed to align more closely with England's plans. (England, Scotland, and Northern Ireland later moved to begin requiring deposits in 2027.)

When it comes to bottle bills, industry has won even where it's lost. For the most part, they've become a way to get containers back for recycling, not a means of returning to the old system of refill and reuse. Companies had managed to "shut down debate over whether disposable beverage containers were a good idea in the first place," one writer observed. Vermont's original 1953 law had "required manufacturers to accept and refill their empties. No one's talking about that now."

CHAPTER 3

America's Plastic Boom

*1961—The Dow Chemical Company unveils its foam packing peanut at a New York plastics exposition.
Ten million metric tons of plastic produced.*

Two tennis courts, a baseball diamond, some old metal picnic tables, and a playground. As small urban green spaces go, Hartman Park, on the eastern edge of Houston, is unremarkable. Except for this: It sits across a narrow street, a hundred meters—if that—from the huge white tanks and twisting pipes of a vast oil refining and chemical complex. Smoke wafts from towers just behind the lone jogger who's doing laps around the park's perimeter the morning I visit, and the dystopian industrial infrastructure stretches far into the distance.

This neighborhood, Manchester, is in the heart of the country's biggest plastic and petrochemical production zone. I've just recovered from an inconveniently timed bout of Covid and have been sleeping badly in a noisy Airbnb, so I'm dragging a little as I get out of my rental car at the park to greet Yvette Arellano, a local activist who's agreed to show me around the area. Arellano is warm and friendly, with long dark hair, round glasses, colorful woven bracelets, and wide, flowing orange trousers, and I begin to perk up as we sit down at a picnic table to talk.

"You assume this is one facility," says Arellano, gesturing across the street. In fact, "there are three facilities here"—an oil refinery, a 101-tank chemical storage complex, and a plant making polyurethane, a plastic used in foam cushioning, shoe soles, and insulation, among other products. Arellano is right that it's hard to tell where one facility ends and the next begins. In some ways, the same is true of this region's fossil fuel infrastructure and the petrochemical industry built atop its foundations. Houston, of course, is well known as the center of the nation's oil and gas industry. Refineries are everywhere. In addition to breaking crude oil down into fuels like gasoline and diesel, they also separate out ingredients that can be used to make all manner of chemicals, including plastic. The Houston area alone is home to more than six hundred refineries and chemical plants.

As fossil fuel companies have leaned hard into plastic, the working-class, mostly Latino and Black residents of Manchester and nearby neighborhoods have watched new petrochemical plants go up and old ones grow. Trucks loaded with pipeline segments or hulking pieces of equipment rumble through the area, and factories' fences are plastered with tacked-up notices of expansion applications.

That rapid growth is part of a boom that's supercharged plastic production all along the Gulf Coast of Texas and Louisiana. The region—more than a thousand kilometers of swampy, sweaty coastline stretching from Corpus Christi, near the Mexican border, through the East Texas refining and chemical hubs of Port Neches and Port Arthur, across the state line to Lake Charles, and up the lower Mississippi River to Baton Rouge—has long been the US petrochemical industry's main home. In recent years, its expansion here—steeper than anywhere but China—has brought a major global realignment. America was already a significant plastic maker, but over the course of the 2010s it became a dominant one.

The roots of this American plastic boom stretch across the country, and deep underground. The source they've tapped is a rich reservoir of chemical ingredients buried within layers of subterranean shale rock. Those raw materials are now flooding to the surface, blasted out by fracking wells, which—as they've transformed the US energy landscape—have also sent plastic production soaring.

Like the plastic boom's roots, its harms reach thousands of kilometers too. Here in Manchester, they come in the form of a toxic brew of chemicals that waft from the plants. Terrible odors often hang in the air, and one team of researchers found dangerous heavy metals including arsenic, lead, and chromium in open drainage ditches. Talking to locals, Yvette Arellano hears stories of nervous system disorders, childhood leukemia, and kids with nosebleeds so bad they need to be picked up from school. One analysis found Manchester residents' risk of both cancer and respiratory illnesses was 22 percent higher than the city average.

And then there are the emergencies—"events," as industry calls them. Explosions and fires rip through plants with frightening regularity, each one spewing out another hit of toxic chemicals. "We're talking about four to six times a year, where all of a sudden you see billowing black smoke" coming from a plant, and no one knows what's in it, says Arellano. "You could be inside doing dishes, going about your business, and all of a sudden you're getting extremely tired, you're getting headaches," nausea, and dizziness. The worst such events bring shelter-in-place orders—with warnings to seal windows with wet towels and plastic sheeting—that can last for days. Sometimes such emergencies are caused by the storms pummeling the region with growing power and frequency. When Hurricane Harvey drenched Houston in 2017, for example, a storage tank's roof failed in the refinery next to Hartman Park, releasing more than one hundred tons of toxic vapors and causing levels of the carcinogen benzene to spike alarmingly in Manchester.

One of the worst episodes came in 2019, when fire broke out at a 242-tank chemical storage farm about fifteen kilometers away. It began when a tank of naphtha—a key ingredient in plastic, although this batch was probably meant to be blended into gasoline—leaked, then erupted in an uncontrollable blaze. Over three days, the fire consumed fourteen other tanks whose contents included xylene, toluene, and pygas, all used in plastics production. Arellano (who uses *they/them* pronouns) was at home, streaming updates online for neighbors who might not know how to protect themselves. When that was done, the terror hit, Arellano tells me, grimacing and slowly shaking their head, then pulling off their glasses to wipe their eyes. At one point, on the phone with a colleague, Arellano recalled, "I told her I was scared, I told her I didn't want to die." By day four, "I was just having mental breakdowns in my closet." One thing Arellano understood that others may not have is that the Houston Ship Channel—eighty kilometers crowded with petrochemical plants and refineries—is terrifyingly vulnerable to a domino-like disaster, triggered by an accident or storm, that could ignite plants along the entire corridor. One local official I spoke to called such an eventuality "catastrophic, unlike anything we've seen in modern history in this country." As Arellano huddled in a closet, the thought felt inescapable: "This could be it. This could be that explosion."

On the playground near our picnic table, a woman sits on the swings beside a small child, then leans over to help the toddler adjust a shoe as Arellano and I get ready to head out for a drive. As soon as we pull away from the park, we're in a neighborhood of modest houses, some neat and well cared for, others boarded up and abandoned. Dotted between them like squares on a checkerboard are empty lots, often behind fences, with signs that say "Valero, Private Property." The company whose refinery I just gazed up at has been buying houses here and knocking them down. For residents, Arellano tells me, this slow hollowing out not only drains the community's vitality but is also depressing property values, making it hard for those who leave to purchase a home elsewhere.

Soon we've left Manchester behind, and I look over it from an elevated stretch of highway. I hadn't realized how close we were to Houston's port. Buffalo Bayou—which in the city's center gives its name to a park complete with running trails and art installations—is crowded here with hulking ships. We pass above a huge rail yard, and as we get onto Highway 225, I gape at yet another complex of dozens of towers and white spherical tanks. The road is lined on both sides with similar facilities, and smoke billows from many of them. This is the Houston Ship Channel, the largest petrochemical complex in the country, and one of the biggest in the world. We're going a hundred and ten kilometers per hour, and the plants—covered with the spaghetti-like tangle of pipes I am learning to recognize as petrochemical production's distinctive look—go on and on, one merging into the next in a relentless, astonishing landscape. Signs bear companies' names, some I recognize and others I don't—INEOS and Invista, Dow and OxyVinyls, Total and CIMA and Chevron Phillips and Mexico's Pemex and Brazil's Braskem. Here and there I see white fumes spraying out of valves, and everywhere flames lick up from high pipes.

We've left Houston proper now, as the Ship Channel cuts through smaller municipalities—Galena Park, Deer Park, Baytown. Amid the industrial infrastructure, we pass homes, hotels, schools with yellow buses out front. Eventually, we leave the highway and enter an even denser production zone, then stop at a railroad crossing. A long train of tanker cars is grinding slowly along the track, its front entering a plant's grounds. The wheels screech as it stops for a moment, then inches backward, then moves forward again. Trains like this—their cars cylindrical, unlike the boxy freight containers I'm used to seeing—are a constant presence in Houston's industrial zones. Their loud horns, sounding all night long, are what's been waking me at my Airbnb. Residents sometimes get stuck at crossings for hours, Arellano tells me, and the danger of an explosion always looms. "It's death" if you're nearby when one happens, Arellano says. Turning around, I see we're boxed in by cars behind us, and starting to feel a little panicky, I look at my watch. The seconds tick by. Later, checking my recording of the drive, I see we sat at the crossing for just eight minutes. It felt a lot longer.

Our tour finished, Arellano leaves me back at Hartman Park, where I get into my rental car and loop over to the side I haven't explored yet. Even after all I've just seen, I'm flabbergasted by the small ranch homes right next to dozens of tall cylindrical tanks—as close as a neighbor's house might be on a tightly packed block. A few chairs sit on a patch of dirt beside one pale purple house, perhaps twenty paces from the edge of the tank farm. I've heard that Houston is the only major American city with no zoning, but until now I hadn't really grasped what that means. The rules that in other places separate residential areas from industrial ones don't exist here. A little farther on, a low metal fence and a strip of grass a meter wide are all that stand between the big tanks and a road barely wide enough for two cars to pass. More houses are right across the street. A German shepherd wanders by, then lies down in the road, and I snap a photo of the dog in front of the line of tanks.

A PICTURE OF HARTMAN PARK HANGS IN LOREN HOPKINS'S SUNNY OFFICE in a municipal building surrounded by hospitals and huge parking lots. Hopkins, the chief environmental science officer at the Houston Health Department, has long blond hair, dangly earrings, and chunky heels, and talks in a mile-a-minute Texas twang. She's in the frustrating position of tracking what she calls the "chemical soup" of pollutants in neighborhoods like Manchester, without having the authority to do much about it. It's the state environment agency—the Texas Commission on Environmental Quality, or TCEQ—that decides what chemicals plants may emit, and holds the power to enforce those limits.

The petrochemical industry's heavy presence makes Houston's pollution problems unique, explains Hopkins, who's also a Rice University statistics professor. The city's air is filled with an ever-changing stew of dozens of dangerous substances. "There's just so many," she says. "It's overwhelming." Particularly worrisome is benzene, which causes leukemia. Scraping data from regulators' air monitors, her team created a system that generates email alerts when levels spike. "We get them every day, all day long," she tells me. The numbers are shocking. "The baseline concentration at a monitor could be 0.25 parts per billion, then we saw, like, forty [parts per billion] for a couple of hours," she says, leaning forward to rest her chin on extended fingers. "It happens again and again at certain locations," she continues. "And that community doesn't know. Nobody's going to do anything about it."

Part of the problem for those living in heavily burdened places like Manchester is that the federal Clean Air Act treats chemical toxics differently than the more common forms of pollution that come mainly from traffic, or non-chemical industries. Toxics are regulated on a plant-by-plant basis, with no overall limit on how much is allowed in the air. "In cities like ours, that just doesn't work very well," Hopkins explains. "We have too many industries next to each other." The state determines what a given plant can emit, "but then the neighbor is doing the same thing."

Seeking to reduce such dangers, the Biden administration tightened enforcement for emissions of chemicals including benzene and 1,3-butadiene in 2024. But it wasn't long before Donald Trump's second administration began ripping up such regulations, part of its broader—and head-snappingly aggressive—assault on rules intended to safeguard Americans' health. Chemical companies were among those lining up to request the loosening of limits on harmful substances their plants put out, but Trump's vision was bigger than any one industry. Impatient with legally mandated processes that make the repeal of any regulation slow and arduous, he began claiming authority to cull them en masse, an assertion of presidential power both sweeping and audacious. In one case, the administration invited companies to send emails to request exemptions from pollution rules, and "the president will make a decision." Like much of Trump's smashing of precedent, it all rested on shaky legal ground, and at the time of writing it remained unclear exactly what would survive court challenges. But with the administration displaying open disregard for established law, and using mass layoffs to destroy the Environmental Protection Agency's ability to enforce whatever rules are left on the books, neither residents of Houston's most polluted neighborhoods nor Americans more widely could expect protection from Washington. The failures predate Trump, but his actions on this, as with so much else, made clear that he would do his utmost to reverse what progress the country had achieved.

For its part, TCEQ, the Texas environment agency, rejects suggestions it allows unhealthy air. The agency told me in an email that while exposure to chemicals is associated with "some level of risk," people encounter many of them "from many sources every day," and "the body can generally remove" harmful ones. Its guidelines, the agency said, reflected "scientifically sound assessments of a chemical's potential for" harm, noting that the vast majority of its many Houston air monitors, including in the Ship Channel, showed annual averages for benzene and 1,3-butadiene within acceptable levels.

A big worry for Hopkins is that Texas deems some chemicals safe at much higher levels than other states do. For 1,3 butadiene, which causes leukemia and other cancers, it regards anything less than nine parts per billion as acceptable. California considers concentrations above one part per billion dangerous. Some Texans may roll their eyes at being compared to liberal California, but companies take advantage of the discrepancy, declining to upgrade plants that would have to be improved elsewhere, Hopkins says. "Same company, different emissions for the same processes."

AMERICA'S PLASTIC BOOM

FRACKING—IT'S SHORT FOR "HYDRAULIC FRACTURING"—WAS A NEW APproach to drilling, able to release tiny pockets of oil and gas from shale rock up to three kilometers deep. Deploying underground explosions, followed by high-pressure blasts of millions of liters of water and millions of tons of sand, the fracking wells that began proliferating across the country around 2005 extracted fuels that had previously been unreachable.

They brought other substances up from beneath the surface too. Chief among them was ethane, its molecules a slightly different arrangement of carbon and hydrogen atoms than those of methane, the main ingredient in natural gas. Unlike its chemical cousin, ethane is not suitable for generating heat or electricity, and at first fracking operators regarded it as waste, sometimes burning it off at wellheads. But petrochemical companies soon saw dollar signs in the torrent of cheap ethane. They began pouring money into plants that could make use of it and other fracking by-products, like butane and propane. In the United States alone, the industry has plowed more than $200 billion into such facilities, turning those chemicals into all manner of products, including paint, detergent, and antifreeze.

And, of course, plastic. There are many ways to produce the multitudinous types of plastic. Central to the huge American expansion are plants known as "ethane crackers." In furnaces as high as fifteen-story buildings, with temperatures surpassing 800 degrees Celsius, they "crack" ethane molecules apart so their components can recombine into the chemical ethylene. A separate process then fuses ethylene molecules together to create polyethylene—the world's most widely used plastic. Similarly, propane can be cracked into propylene, to be processed into polypropylene, another versatile and common plastic, used to make shopping bags, diapers, and backpacks, among other everyday items. A variety of other plastics—such as nylon and polyurethane, whose innumerable applications include toothbrush bristles, umbrellas, carpeting, and car bumpers—are also produced from fracked chemicals. The ingredients' rock-bottom prices and ubiquity turned the United States into the cheapest place to make plastic.

Trey Hamblet had never seen anything like it in his decades watching the chemical industry. He's a slim man in black jeans, a big silver belt buckle, and cowboy boots with pointy toes. His cowboy hat, he tells me with a laugh, is outside in his truck. Hamblet's office is in a nondescript building in Sugar Land, a suburb of eight-lane roads and huge malls just outside Houston, about a half-hour drive from Hartman Park, where I'd met Yvette Arellano. Being originally from New Jersey, I didn't think such developments could surprise me, but I'm a bit taken aback by the Super Target nearby, since I thought regular Targets were already supersize. Industrial Info Resources, where Hamblet is a vice president, tracks plant construction and sells its data to those who invest in and work with the industry. "We're physically—old school—picking up the telephone" to call site managers and engineers, he explains, amassing data so granular he can identify a petrochemical plant's individual boilers. "Every plastics unit that's been built for the last twenty years, I can tell you down to the month when it started up."

The frenzy began in 2013, he recalls. Before fracking, American plastic makers' main ingredient was naphtha, which is derived from crude oil. It looked to be growing scarcer and more expensive in the United States, making it harder for US producers to compete globally. Texas plastic plants were closing, and as one new one broke ground, "I thought 'Gosh, this'll be the last billion-dollar project I see in my career,'" Hamblet says. "Roll the clock forward, and I've got dozens upon dozens." The driving force was obvious. "We almost overnight had access to ethane" that was previously unavailable, he tells me, "an incredibly abundant, inexpensive feedstock." (*Feedstock* is the industry term for the ingredients that go into chemicals such as plastic.)

For Hamblet, even keeping tabs on it all was sometimes challenging. "To say we were busy—yes, that's an understatement. It was a lot of action going on simultaneously," he says, laughing. At one point, he recalls, there were seven major North American ethane crackers under construction. "We had at no point in our past ever had more than one" being built at a time. The industry itself was as surprised as anyone: "It's absolutely extraordinary this is happening in the United States," an ExxonMobil executive told the *Houston Chronicle*. "Nobody predicted this." Fracking was the only reason.

The boom may have happened in the United States, but the participants are not only American. The Gulf Coast is dotted with plants built by companies from Saudi Arabia, Qatar, Taiwan, France, and Austria, among other nations. US operators are in on the action too. ExxonMobil touts its $20 billion expansion in the region with the slogan "Growing the Gulf." Its projects include a $10 billion plant near Corpus Christi, jointly owned with the Saudi firm SABIC, that is said to be among the world's largest ethane crackers, as well as a big expansion at a facility in Beaumont, east of Houston, that will boost polyethylene production there by 65 percent, to 1.7 million tons annually. Just a year after firing up a new cracker in Baytown, on the Houston Ship Channel, ExxonMobil said it would put $2 billion more into that complex. Chevron Phillips has its own mega-plant nearby, plus two more a short drive away.

Around 2019, Hamblet tells me, a global glut of raw plastic was accumulating, and the construction spree began to slow. The Covid pandemic brought more disruption, but industry remains confident demand will eventually catch up to the new supply it's pumping out. So while the 2020s are bringing less building than the 2010s, new projects are still going forward. Hamblet doesn't see that changing any time soon. "We're going to absolutely require lots more plastic," he says, "and it will always be built in the cheapest place first." Given the continued flow of fracked ethane, that, for the foreseeable future, is North America.

And of course the plants that have already been built will be churning out plastic for decades to come. The numbers are jaw-dropping. In one five-year period, North America's virgin plastic production jumped by 60 percent. By 2022, upwards of 680 million kilograms of raw plastic a month—more than 1,300 containers-full a day—was being exported through Houston's port alone, an amount that says nothing of the material transported domestically by train and truck.

Plastic makers are reluctant to speak to me—the companies and Gulf Coast–based trade associations I reached out to mostly declined or ignored my requests to meet. But they talk to Hamblet. When I ask him why they're so confident the world will want the ever-increasing amounts of plastic they're churning out, he says the answer is easy: a growing global population, buying more and more stuff. In particular, producers see big potential in the growing middle classes of China, India, Southeast Asia, and beyond, where hunger for new appliances, electronics, cars, and other goods is rising, Hamblet says. "You can't have any of them without chemicals or plastics." Indeed, the industry says as much publicly. "Those overseas markets are the motivation behind our investments," ExxonMobil chief executive Darren Woods explains on the company's website. "The supply is here; the demand is there. We want to keep connecting those dots."

THE INDUSTRY'S WEALTH AND ECONOMIC HEFT HAVE GIVEN IT A GREAT deal of sway with Gulf Coast political leaders—who have, in turn, made the region a comfortable home for petrochemical companies. During almost two decades as chief toxicologist at the environmental regulator TCEQ, Michael Honeycutt was the man steering Texas's decisions on how much chemical pollution is too much. Over the years, Honeycutt, who has a boyish face and brown hair combed to one side, shocked some fellow scientists by expressing views well outside the mainstream. Testifying in Congress, he sought to poke holes in widely accepted research showing sooty air pollution particles shorten lives, instead suggesting they might help people live longer. Another time, he questioned the long-held consensus that mercury, often ingested via seafood, is a harmful neurotoxin, saying Japanese people eat more fish than Americans and have higher blood mercury levels yet do better on IQ tests.

In his years at the agency, Honeycutt led assessments of acceptable exposure levels for dozens of chemicals, each one underpinning subsequent decisions on what plants would be allowed to emit. Two-thirds of his office's assessments relaxed limits, even though many were already among the loosest in the nation, a 2014 analysis found. One expert told *Inside Climate News* and the Center for Public Integrity that the Texas agency's decision to increase by 40 percent the concentration of benzene it deemed safe was "the most irresponsible action I've heard of in my life."

Honeycutt had long been close to industry, even giving the keynote speech at a conference of the American Chemistry Council, the companies' powerful trade group. When he and his team began reevaluating limits on ethylene oxide, a chemical used to make antifreeze and detergent, as well as plastics, and which is linked to breast cancer, leukemia, and lymphoma, they met and spoke with representatives of big chemical makers including Shell, Dow, and BASF, as well as the council itself. The agency and the companies said at the time there was nothing untoward about their interactions, and denied suggestions the reassessment had drawn on a draft provided by industry. But companies got what they wanted. Honeycutt's agency proposed an acceptable ethylene oxide exposure level fifty times looser than its own previous assessment—and 3,500 times weaker than federal guidelines recommended at the time. "They're just low-balling the whole thing," one expert told news site *The Intercept*. The agency declined my request to interview Honeycutt, who has since retired.

Texas's permissiveness, of course, is far bigger than one official.

"We're in a state that has set every single rule of the road in favor of industry," says Christian Menefee, who, as Harris County attorney, is in charge of civil litigation for the county that includes Houston. "It feels at times like it's David and Goliath." While conservatives traditionally (although selectively) have purported to reject federal interference in state matters, Texas's hard-right political leaders are happy to meddle in local affairs on fossil fuel and petrochemical companies' behalf, he tells me at his downtown office, where a drawing of the late Supreme Court justice and civil rights icon Thurgood Marshall sits on a shelf above rows of law books. A Democrat who combines a politician's charisma with a lawyer's precision, Menefee was just thirty-two when he was first elected in 2020, the youngest ever, and the first Black person, to hold the job. A year earlier, the then-Speaker of the Texas House of Representatives had been caught saying he wanted to make the upcoming legislative session the worst ever for cities and counties. In the months that followed, Menefee recalls, "bill after bill was proposed" to reduce local officials' authority. An earlier law already allowed the state to seize control of municipalities' lawsuits. So industry-friendly state lawyers can wrest a case from him "and settle it for pennies on the dollar. It's happened time and time and time again," he tells me. The legislature "is continually tying our hands behind our back."

County officials only intervene to begin with because the state environment agency is not doing its job, Menefee says. After years of being shaped by Texas's conservative politics, it doesn't see itself "as a regulatory body or enforcing authority," he believes. "I think they view their role as a facilitator. Like, 'We are facilitating you getting your permits,'" or polluting the air, he says. "As opposed to 'We're reviewing your permit application to see if you're going to be harmful to communities,'" or "'We're going to court if necessary to enforce the laws.'" (The agency, in its email to me, declined to comment on the accusation of pro-industry bias.)

Menefee is well aware of the fossil fuel and petrochemical industry's economic importance for Houston, where he's seen it provide opportunities to friends, neighbors, and his own father. "There are a lot of good people who work in the industry," and he's not trying to shut it down or seeking unreasonable limitations, he says. "I'm all about fairness. There are rules of the road. If you run a stop sign and a police officer's around, you're going to get in trouble. We have companies routinely running stop signs in this area, and we have a state regulatory agency that is asleep at the wheel" when it's not actively thwarting enforcement by others. "In this country, we hold people accountable for all kinds of things. But in Texas, we don't often hold industry accountable for their mistakes."

TEXAS DOES PLENTY OF ITS OWN DRILLING. BUT THE RAW MATERIALS FEEDing its plastic production come from farther afield too. The Marcellus Shale formation, stretching from upstate New York all the way to Tennessee, is the world's second-largest methane (or natural gas) field, and it's also rich in ethane. The bulk of the reserves lies beneath the upper Ohio River Valley—the northern swath of Appalachia, comprising western Pennsylvania, northeast Ohio, and West Virginia. The region was once one of the country's great industrial corridors, but with the loss of much of its steel and coal industries, it hit hard times as the twentieth century waned.

Since the mid-2000s, it's been fracking country. The ethane flowing from its wells drew plastic makers' eyes. Shell decided in 2016 to build a $6 billion cracking complex northwest of Pittsburgh to turn fracked Appalachian ethane into 1.6 million tons of raw plastic a year. The state handed over a $1.6 billion tax break—the biggest in Pennsylvania history—to help executives make up their minds.

Nearby Washington County, just south of Pittsburgh, is the state's most heavily fracked, home to more than 1,500 wells. Lois Bower-Bjornson's family has lived there for two hundred years. The white clapboard house she and her husband are raising their four kids in was once the local general store, and it sits on five woodsy hectares. The area is bucolic and beautiful—farmhouses dot its gentle hills, and after a long day of rain, the green trees and fields twinkle in the late afternoon sun as I arrive. Bower-Bjornson—tall and lanky, with dark jeans and low brown boots—is emptying Goldfish crackers from her daughter's lunch box, and her dog wanders across the kitchen to nuzzle me hello. A few minutes later, we're sitting in an old barn converted into a studio where she teaches dance and yoga.

Life here isn't as peaceful as it looks, she tells me. I've seen hints of that already, in the tanker trucks roaring down her narrow road. I've been in western Pennsylvania long enough to know they carry fracking wastewater—a toxic mix of some of the one thousand different chemicals used in drilling, plus naturally occurring radioactive elements unearthed from deep underground. A single well can spew more than fourteen million liters. And just a few minutes from Bower-Bjornson's house, I passed the red metal tower of a drilling site. "It's all around us," she says.

I've come here to follow the forces driving the American plastic industry's huge growth spurt to their source. Fracking in Pennsylvania—and its neighbors West Virginia and Ohio—supplies the nearby Shell cracker, but the wells Bower-Bjornson's family lives among are also connected by a web of pipelines and international shipping networks to the plastic plants I drove past outside Houston, and some even farther away, in Europe and Asia. So the drilling that happens here—and the harms that come with it—are inextricably linked to the rising tide of plastic in all our lives.

Bower-Bjornson and her husband moved back to Washington County from Pittsburgh in the mid-2000s, after spotting a For Sale sign on their way to visit her parents. Their rambling old home quickly became a holiday gathering place for friends and family. But there were other visitors too. Not long after the family's arrival, landmen came knocking—representatives of fracking companies like

Chevron and Range Resources who were blanketing the area, offering money for leases allowing drilling, pipelines, and such on homeowners' properties. Back then, the industry was in wooing mode—handing out "Friendly Fracosaurus" coloring books, funding renovations at the county fairground, showering donations on local fire departments and charities.

As fracking operations proliferated, unnerving illnesses soon followed. All four of her kids have had health problems, Bower-Bjornson tells me. The oldest got a rash all over his body. Her daughter had eczema so severe that if she hadn't been the one bathing the girl, "I would have thought someone scalded her," and her youngest son, Gunnar, suffers severe nosebleeds and other mysterious symptoms. Bower-Bjornson enrolled the family in a study in which their urine was tested for signs of exposure to some of the one-hundred-plus chemicals known to pollute air around frack sites. Gunnar's numbers were "off-the-charts high. All the kids are high, but his are the worst," she says. Researchers also tested the family's drinking water and found twenty chemicals commonly used in fracking.

Others have suffered far worse. In 2016, Bower-Bjornson's friend Janice Blanock lost her nineteen-year-old son, Luke, to Ewing's sarcoma, a rare bone cancer that afflicts about one in a million Americans annually. Between 2008 and 2018, six people living in Luke's school district, Canon-McMillan, got it. The *Pittsburgh Post-Gazette* documented at least twenty-seven diagnoses in a decade across four counties, including Washington, with a combined population of about 750,000. Had the region's incidence hewn to national levels, there would have been eight. It's not just Ewing's. "There's cancer in every single school here," Bower-Bjornson says. "Rare cancers."

A state-funded study eventually found no association between Ewing's and fracking, and, to be sure, there are other sources of industrial pollution nearby. But the analysis concluded children living within 1.6 kilometers of a well were five to seven times more likely to get lymphoma than kids more than eight kilometers from one. Yale researchers found children living near frack wells at birth were two to three times more likely to get leukemia, and a Harvard team discovered elderly people living near fracking sites were at increased risk of early death. The industry, saying its operations are safe, slammed the studies as "efforts to advance an anti–natural gas agenda, drive more dollars to already well-funded activist organizations, and of course— serve as internet click-bait." Other research has tied proximity to fracking to elevated rates of heart failure, breathing and skin problems, migraines, and hospitalization for cardiovascular and nervous system ailments, as well as miscarriage, premature birth, and congenital disabilities. On top of illness, those scientists noted, fracking also fuels climate change—largely because drilling, processing, and transporting gas leaks a great deal of methane, which is a powerful driver of warming.

Researchers estimate it takes one thousand new wells to keep a cracker the size of Shell's supplied with ethane. So it's not just that fracking enables plastic production. The relationship goes both ways. "You have to drill the wells to support the petrochemical plant, but you also have to build the petrochemical plant in order to keep drilling the wells," one expert told *Environmental Health News*. Fracking is built on shaky economic ground, its profits seesawing along with ever-volatile oil and gas prices. The extra revenue stream ethane generates is a valuable financial cushion. In hard times, it can mean the difference between keeping wells pumping and shutting them down. In flusher years, ethane sales pad healthy profits further. With much of that ethane eventually becoming single-use plastic, critics coined a term to highlight the connection: *frackaging*.

PIPELINES ARE KEY TO LINKING FRACKING WELLS TO THE PLASTIC PLANTS they feed—tens of thousands of kilometers of pipelines crisscross Pennsylvania alone. So the hazards they pose are part of plastic's story too. Karen

Gdula came to understand that gradually, after a disastrous day that brought the dangers frighteningly close. Gdula lives in the house she grew up in, a modest home on a pretty street northwest of Pittsburgh. Ivy Lane, she tells me, is someplace special. "There's a warmth and a caring," she says, sitting across from me at a table just off her pink-carpeted living room, knickknacks on display in glass-fronted cabinets. "We look out for each other."

The street's neighbors never needed those bonds more than on September 10, 2018—and in the weeks and months that followed. Retired and newly married, she was asleep when, just before 5 a.m., an explosion shook her home, recalls Gdula, who has shoulder-length blond hair, wears a floral patterned shirt with tan slacks, and speaks in matter-of-fact Midwestern tones. Some of her neighbors thought it was a plane crash, but when she and her husband saw a fireball stretching above the tops of the towering pine trees across the street, they knew exactly what had happened. The Revolution Pipeline, running right behind Ivy Lane, had come into service only days before. "A regular fire doesn't get that big that fast," she knew. It had to be the pipeline.

Gdula and her husband threw together some clothes and medications. As they drove away, she tells me, tears in her eyes, "I said, 'Goodbye, house. I don't know if I'll ever see you again.'" Luckily, the woods were sodden after several days of rain, and the wind reversed direction, pushing the blaze back into an area that had already burned. Otherwise, "it could have come up and wiped out this whole street," she says, flipping the pages of a white binder she's filled with photos of the blast's aftermath. No one was hurt, but the explosion flattened a home three doors down—in Gdula's scrapbook, it's a smoking foundation surrounded by charred trees—and toppled six big electrical towers. It was one of the worst oil and gas accidents in recent Pennsylvania history.

Gdula and her husband were able to return later that day, but life on Ivy Lane was anything but normal. Heavy equipment was soon rolling down the narrow street as workers began the difficult process of stabilizing the steep slope where land slippages had ruptured the pipeline. But what really bothered Gdula was the news that another pipeline was coming to her street. Known as Line N, it would supply gas to power the new Shell ethane cracker, then under construction, just five kilometers away (it began production in 2022). Today, Revolution—rerouted to avoid cutting across the hill—runs even closer to Gdula's home than before. Unlike Line N, it is not linked directly to Shell's cracker, but it, too, is likely supplying plastic production. (Shell declined my request for an interview, its spokesman citing "fatigue for the stories that continually draw unflattering conclusions.")

Gdula spent most of her career at a company that designed equipment for oil and gas workers, so she wasn't someone inclined to distrust the industry. Before the explosion, she'd been reluctant to criticize fracking's spread across her state, because, she says, "I like to be able to turn on my lights, I like to have air-conditioning." But it was disconcerting to realize how much of the drilling was for plastic production, rather than for power plants or home heating.

Nearly four years after the explosion, Revolution's operator—a subsidiary of pipeline giant Energy Transfer—was convicted on criminal environmental charges. The state attorney general said the company had "repeatedly ignored" protocols for preventing landslides and failed to adequately secure the pipeline into bedrock. And Revolution is not Pennsylvania's only troubled pipeline. Energy Transfer was also convicted in spills of thousands of liters of drilling fluid from its Mariner East II line, which carries ethane to a facility near Philadelphia, where it is loaded onto ships for plastic makers overseas. Similar problems plagued construction of the Falcon pipeline, which feeds ethane to Shell's plant. Underground pipelines are marked by low poles with color-coded tops, and when Gdula takes me for a drive, she points some out in front yards, just ten meters from homes. Later, winding along a creek in a nature preserve a few minutes from the Shell plant—the kind of place I'd happily visit for a weekend hike—I see them dotted around the woods.

Then there are the trains. About thirty kilometers from Ivy Lane, a train carrying about 400,000 liters of dangerous chemicals derailed in East Palestine, Ohio, on the Pennsylvania border, in February 2023, making national news as black smoke billowed from its wrecked cars. Among the most concerning cargo was vinyl chloride, which is linked to brain, liver, and lung cancers, as well as lymphoma and leukemia, and is used to make polyvinyl chloride, or PVC plastic. The train had been heading from a plant on the Houston Ship Channel to a New Jersey facility that makes vinyl flooring. It was plain to see, Gdula tells me, that emergency responders were ill-equipped for such a complex disaster. With similar trains coming and going from Shell's cracker—not to mention the operations of the plant itself—she's not taking any chances. "My husband and I have already talked about it," she says. If something goes wrong at Shell, they're not waiting for an evacuation order. "We are getting in our vehicle, and we are leaving."

THE 140-KILOMETER CORRIDOR THAT FOLLOWS THE LOWER MISSISSIPPI River from New Orleans to Baton Rouge is crammed with 150 petrochemical plants and refineries, and their pollution has saddled the region with the unenviable nickname "Cancer Alley." The river twists and turns, bending back on itself as it winds west, and then north. I can't see the water as I drive alongside it—it's hidden behind an earthen levee that rises from the road as a low grassy hill. On my other side, sugarcane fields stretch far into the distance, cut through here and there by an isolated street or two of homes. Hulking industrial facilities are dotted among the fields too, and from the long bridges that arch high above the Mississippi, I see plant after plant, each on its own big rectangle of land. Many are connected to the river by pipes and chutes that cross above the road and over the levees. It's a rural version of the infrastructure I glimpsed along the Houston Ship Channel, and amid it all, a sense of emptiness—an abandoned sort of quiet—hangs over much of the area.

Tourists drive this stretch of the Mississippi to get a taste of antebellum history at well-preserved plantations, some of which even host weddings. The Whitney Plantation is different. It's the only one focused entirely on the people whose forced labor created the wealth that once gave this region the nation's highest concentration of millionaires. Walking through its grounds is a gut punch—displays offer an unvarnished exploration of slavery's brutal, agonizing history here. Memorial walls list the names of those enslaved on the plantation, often only first names, with birth years and occupations. Narrators read harrowing testimony of physical and sexual violence, family separation, emotional anguish. There's a glimpse of slavery's afterlife too—the poverty that trapped cane workers long into the twentieth century, as they became sharecroppers and then employees.

A straight line connects that dark history with the region's present. Many of those living on the lower Mississippi today are descended from the enslaved, and the slave and sugar economy shaped the very geography of this place. After emancipation, Black people formed "free towns" on strips of land around plantations' edges. The bulk of an estate typically remained in one large parcel. After oil was discovered in Louisiana in 1901, refineries began buying those plots, and petrochemical companies soon followed. Like the plantation owners, they valued the access the river provided to global markets. Plus, there was no need to hassle over acquiring and consolidating multiple properties. "You can get a thousand acres [four hundred hectares] from one person, because they're a descendant of a plantation owner," one local activist explained to me. While white owners pocketed millions, Black communities got nothing. Now, "instead of plantations," an *Atlantic* writer put it, "Louisiana's historic free towns share fence lines with plants."

Over the years, some companies have responded to particularly damning evidence of pollution by buying out affected communities, whose residents, with few other options, took what they could get and scattered. It's part of why the place feels so empty. "You don't have a grocery store, you don't have a gas station, you don't have a bank," one lifelong resident of St. James Parish tells me. "I see how the community has just diminished." St. James High School moved from its longtime home to make way for the offices of Koch Methanol, whose nearby plant turns fracked chemicals into ingredients for products including paint, adhesives, and plastics.

In Robert Taylor's neighborhood, a grid of neatly kept homes on small patches of land a half hour west of New Orleans, kids are playing in yards and adults potter around doing weekend chores on the sunny Saturday I visit. Taylor has been living alongside the plastic industry since well before its most recent spate of growth. The chemical giant DuPont began building its plant in St. John the Baptist Parish, as Taylor recalls, in 1963, the year he got married. The complex started producing neoprene—synthetic rubber used in goods from wetsuits and laptop cases to conveyor belts and electrical insulation—in 1969, shortly after his youngest daughter was born. He's in his eighties now, in worn dress shoes and a blue blazer that hangs off his thin frame, with white hair and a thick salt-and-pepper mustache. It's been a year and half since a hurricane all but destroyed his home, but repairs are still underway, so he's living in a trailer out front with his middle-age son and teenage granddaughter. As I perch on its front step, he pulls up a chair and sits across from me.

Taylor grew up in St. John parish (parishes are Louisiana's equivalent of counties), on land owned by the sugar company his father worked for. As a young man, he played the Hammond organ and bass guitar at nightclubs up and down the river and on Bourbon Street in New Orleans. "I made my living entertaining white people," he recalls, but when civil rights laws stopped venues from putting Black musicians in segregated spaces before and after shows, some chose to cancel gigs rather than let them mingle with whites. "My primary source was a club out on the bayou here. I'll never forget the lady calling me up, saying, 'Bobby, I'm sorry, but you're not going to be able to play in my place anymore.'" So he trained as an electrician, and later became a general contractor.

The home he and his wife built in tiny Reserve is just a few blocks from the neoprene plant, a vast complex of tangled pipes, big tanks, and towering smokestacks—so close Taylor could sometimes hear announcements from company loudspeakers. As his children grew, they'd complain of chest pains while playing outside, and he sometimes noticed bad smells late at night. Illness was everywhere. Gesturing up the road, he points to two homes where both parents and at least one adult child endured cancer. Another neighbor died from the disease, and two of that man's children now have it, Taylor tells me. "We can go up and down this community, and you will not find a family that has not been touched." That includes his own. "I have a thirty-five-year-old nephew right now that's just been diagnosed" with prostate cancer, he says. "His uncle before him died from it. My brother died of lung cancer. My sister. My mom died of bone cancer. Uncle Johnny died of prostate cancer, and Johnny Jr. did. And my favorite cousin, Dorothy."

Taylor's wife got cancer too. She survived, but struggled with the aftereffects, and moved to California to escape the plant's pollution. Taylor wanted to join her, but she insisted he stay to care for their daughter, who's debilitated by an autoimmune disorder. "She didn't want to leave our baby here, but Raven can't travel at all," Taylor tells me. Raven's condition is extremely rare, but several other women in town have been struck by it too, he says.

In 2015, DuPont sold its plant to Denka Performance Elastomer, a Japanese company. The following year, Robert Taylor finally learned why so much sickness surrounded him. At a meeting with residents, an official from the national Environmental Protection Agency said their neighborhood had the country's highest risk of cancer from airborne chemicals—a risk later estimated at almost fifty times the national average. For decades, the people of Reserve had been breathing sky-high levels of chloroprene, the main ingredient in neoprene. In 2010, the agency had concluded the chemical was much more dangerous than previously thought, deeming it a likely carcinogen that damages DNA. Airborne concentrations around the plant were averaging up to ten times the agency's maximum acceptable limit. Longtime residents had suffered even greater exposure—in the 1980s, the plant was emitting five times as much chloroprene as it did by the latter half of the 2010s. And chloroprene was just one of forty-five dangerous pollutants in the parish's air. "It was mind-blowing to me," Taylor said. "We just had no idea of what they were doing to us. But they knew."

He helped found Concerned Citizens of St. John and started holding meetings at a local church. Referencing national regulators' estimate of 0.2 micrograms per cubic meter as the maximum acceptable chloroprene level, the group got the slogan "Only 0.2 will do" printed on red T-shirts. "Most people don't know what a microgram is," Taylor says, his gnarled hands, with long musician's fingers, folded in his lap. He tells them a gram is about a quarter of a teaspoon. "Divide it into a million pieces" and just a fifth of one is safe in a cube of air a meter on each side. In fact, Environmental Protection Agency scientists said that while 0.2 was the upper limit, they prefer chloroprene concentrations to be one hundred times lower. Such numbers, Taylor says, show "how dangerous that stuff is."

At first, he steered clear of talking about race, but it's so central to his community's experience—93 percent of those living within a kilometer and a half of the Denka facility are Black—that it ultimately felt unavoidable. "The further you get away from the plant in any direction, the whiter the population," he says. "We were termed the low-hanging fruit," unable to leave, or to stop the pollution. There are other plants nearby too: the country's third-largest oil refinery and a chemical maker emitting ethylene oxide—the carcinogen Michael Honeycutt helped ease rules for in Texas.

Denka didn't reply to my emails but has said there's no evidence of increased illness near its plant, and notes it's cut emissions by 85 percent under a deal with Louisiana regulators. The company "is focused on being a good neighbor" and "will continue to look for ways to improve," its spokesman said. Under President Biden, the Environmental Protection Agency acknowledged the reduction but said "there is no question" those near the plant still faced an "elevated cancer risk." Like its other efforts to tighten regulation of the chemical industry, the Biden administration's push to reduce toxic exposures in St. John parish was short-lived. Six weeks after taking office, the Trump administration—alongside its wider smashing of regulations and the federal government's capacity for enforcing them—dropped a lawsuit Biden's team had filed as part of an effort to force Denka to reduce pollution, slamming it as "radical DEI," or diversity, equity, and inclusion.

It's obvious to Taylor that his neighborhood was deemed a "sacrifice zone," the phrase activists use for communities consigned to live with pollution for the sake of someone else's prosperity. "I still haven't been able to wrap my mind around that and understand who had the authority to decide that we were expendable," he tells me, the sadness in his voice mixed with fury. Most of the plant's 235 workers live outside the area. "We're in a parish with billions of dollars of industry, and we're the poorest people in the country," he says. "To kill the people in Reserve for two hundred jobs—I mean, I don't know if anybody's lives and health is worth somebody's job."

CHAPTER 4

"Guilt Eraser"

1967—The Graduate gives plastic its immortal moment on (celluloid) film: "Just one word," an acquaintance advises Dustin Hoffman's character. "Plastics." Twenty-five million metric tons of plastic produced.

Gary Anderson was studying architecture at the University of Southern California when he saw the poster. It was 1970, and environmentalism was in the air, along with a broader antiestablishment vibe. "It was hippies, it was love-ins, it was be-ins," he recalls. "A big stew pot full of ideas and emotions." He shared those ideals. Plus, he didn't have much money, so the $2,500 prize on offer may have been part of why the ad caught his eye too. The Container Corporation of America—a major cardboard box maker—wanted a symbol to represent recycling, and it was sponsoring a contest to find one.

Anderson had studied some graphic design, and he had an idea of what a good logo should aim for: "It needed to be very clear, very simple. And so I thought, 'Well, I could do this,'" he tells me. "'I already have the drafting tools.'" He sat down to brainstorm. The word *recycle* "brought to mind something kind of circular, in motion," reminding him of a long-ago school trip to a newspaper printing plant. He'd been entranced by the big rolls of paper and the long sheets of it running through the air, from one part of the press to another. He thought, too, of his fascination with the Möbius strip—a loop of paper with a twist in it that has unique mathematical properties and carries a sense of the infinite. And he pictured the coffee table tome he'd recently leafed through at a bookstore, full of images by M. C. Escher, whose work often makes use of the Möbius strip, and in which "space just seems to be turned inside out."

Anderson wanted to bring all those things together. He'd been working with arrows in another project, and that was probably why he thought to include them. Over a couple of days, he went through different iterations, and in the end he submitted three versions. It was the simplest one, arresting in its elegance, that got the prize. Anderson had arranged three arrows in the shape of a triangle—each bent over itself, so a viewer seemed to see both sides of a flat strip—all chasing one another in an endless loop. With heavy black glasses and a pen poking from his shirt pocket, he posed for a photo with a Container Corp. executive. The company flew him to its headquarters, in Chicago, then to Aspen, Colorado, for a conference. Anderson shrugged when saw his design soon after on a bank statement, thinking its moment would "maybe last a season or two, and then you'll never hear about it again." For a while, that's what seemed to have happened. He used the prize money to study in Sweden, and later got a job in Saudi Arabia. Sometime in the early '80s, he wandered into an Amsterdam square during a stopover on his way to the United States. There were several big recycling bins, and his design was plastered onto each one—as big as a beach ball, he recalls, holding his hands wide. "I thought, 'Well, it must have caught on,'" since it was being used so far from home.

The three chasing arrows only grew more visible. Whenever Anderson saw them, he'd check whether they'd been reproduced correctly; sometimes, the arrows were flattened, eliminating the multidimensional feel. But mostly, he and his creation parted ways. His résumé didn't mention he'd designed one of the world's most recognizable symbols until late in his career. Applying for architecture jobs, he always worried a potential employer might think, "Well, we're not after a graphics designer." For years, even many of his friends didn't know. "How do you drop that into" a conversation, he wonders. "I didn't want to seem like I was bragging."

Now in his seventies, a Baltimore retiree, Anderson thinks his younger self "would be astonished" by the logo's journey. He knows "recycling can kind of be a crutch sometimes," making us feel better about all the waste in our lives. But it's still better to recycle than not, he says, and "it makes some kind of a dent in this unbridled consumerism that we seem to have developed." He implores me not to turn his words into something negative, because that's not what the chasing arrows are to him. All these years later, he says, "I just look at that symbol and it makes me feel good."

Rightly so. It's a perfect visual, conveying just what Anderson intended: the idea that materials could be repurposed again and again in an endless loop. Infinity, it hinted, could exist within the constraints of finite space—and, by extension, finite resources. That's a seductive thought. So while Gary Anderson had conceived it for paper and cardboard, it's not surprising, in retrospect, that the plastic industry saw something valuable in his design. Soon, companies would grab hold of the image—and the idea of recycling—and warp them beyond all recognition.

RECYCLING IS, AS AUTHOR OLIVER FRANKLIN-WALLIS PUTS IT IN *WASTELAND*, "as old as thrift." Rags were once pulped to make paper, household ashes went into cement, and in Victorian-era London "rag-and-bone men" knocked on doors collecting scraps to sell. "For almost all of human existence, raw materials were expensive, rare, and precious, and so objects would have many lives," Franklin-Wallis writes. Recycling's modern iteration, beginning in the late 1960s, was something different. No longer motivated primarily by the need to salvage hard-to-come-by materials, it had morphed into a tool for managing unprecedented amounts of trash.

Even as industry pushed to frame that ever-growing waste mountain as simply a disposal problem, companies could see that their push to defuse the public's worries about it would have to run on more than one track if they wanted to prevent "well-meaning but misinformed authorities" from stepping in with their own solutions, as the magazine *Modern Plastics* put it in 1966.

With the environmental movement gaining steam, a promising new tool presented itself. As the 1970s dawned, a handful of municipalities were launching curbside recycling collection, typically for newspapers, cans, or glass bottles; by the middle of the decade more than 130 American towns and cities offered the service. "When it started, recycling was a pretty radical idea," one expert told me. To many, it looked like "the only ecologically sensible long-term solution" for a nation "knee-deep in garbage."

Soon industry was trumpeting its own commitment to recycling. In 1971, Coke's New York bottling company sponsored seventeen collection centers in and around the city. Like anti-littering campaigns, such efforts often wielded the language of personal responsibility. Praising schoolkids' participation, one of the program's spokesmen scolded their parents: "If adults would learn to be as conscientious as children, maybe we wouldn't have the problem we have with discarded cans." Coke, along with Pepsi, Mobil, and Reynolds Metals, plastered city buses with ads promoting their work on recycling. "The plastics industry is at work on a number of projects designed to turn waste into something useful," a 1973 Mobil campaign announced, citing a project to mix scrap plastic into concrete, which it promised would result in material "as strong as conventional concrete," but lighter. (A bridge built with the stuff eventually collapsed into a river.)

Plastic wasn't the focus of most early recycling. But even though aluminum, glass, and paper waste—unlike plastic—had decent resale value, a problem quickly became apparent. "Recycling so far is not paying its own way," *The New York Times* reported. Proceeds weren't enough to cover costs, and while big companies were happy to sponsor some eye-catching trials, they didn't want to pick up the tab for long. By 1978, nearly all the three thousand drop-off centers opened in the initial burst of enthusiasm had closed.

Modern recycling's foundational question was now on the table: Who would pay for it? Companies had an answer: government. And since municipalities often lacked the money for expensive new sorting and processing plants, industry pressed Washington to pony up. Executives from Coke, Anheuser-Busch, and the American Can Company testified in favor of a 1970 law authorizing federal grants for local recycling. Six years later, again with big beverage companies' backing, such support was expanded. The idea that this worthy new endeavor would be a public responsibility, not a corporate one, was becoming entrenched. And like so many of the principles industry incubated in the United States, that notion would quickly be exported. Not everyone bought it. Taxpayers, one environmentalist argued, shouldn't be on the hook for dealing with "materials that need not have been produced in the first place." What's more, she wrote, recycling had "a frightening potential for institutionalizing waste generation." Indeed, that was exactly what industry liked about it.

EARLY ON—LIGHTWEIGHT CONCRETE ASIDE—PLASTICS COMPANIES MOSTLY didn't envision their product as part of the new era of recycling. They contended instead that it was well suited for landfills, and even helped keep them stable. Bags in particular, Mobil claimed in one ad, "offer environmental advantages when disposed of in dumps and landfills." But landfills—subject to new regulations on toxic runoff, and pushed farther from population centers by suburban sprawl—had gotten expensive.

In any case, consumers didn't seem keen on having their plastic trash buried. Companies had another idea. Perhaps burning it would be more appealing—especially if it could be painted as environmentally friendly. "Recycle plastic packaging? An excellent idea. But let's recycle it into energy" via incineration, one packaging executive suggested in 1971. That framing persists today, with incineration often branded under green-sounding names like "waste to energy" that imply it's a virtuous win-win. (Near my home in London, I often see trucks emblazoned with slogans like "Powering the Circular Economy" headed toward a big local trash-burning plant.) Sometimes incinerators even win government subsidies as a form of renewable power.

In reality, even high-tech, modern incinerators can spew dioxins and other harmful pollutants. And since plastics are made from oil and gas, burning them delivers a wallop of climate-warming carbon dioxide. While those dangers didn't stop industry from pushing it, there was a bigger obstacle. "The public almost universally believes recycling is the answer," said an executive from B. F. Goodrich, a big vinyl producer. Industry, he argued, had to take heed. Plastics' markets were "under attack," and negative views of it "our major challenge," he explained. "Recycling is a key part of the perception issue."

Perception, in this case, mattered more than reality. So as the tide of plastic rose, companies' recycling boosterism expanded to embrace it. "Recycling sounds like an ideal solution," cheered Mobil. Publicly, it said even plastic bags could be recycled, while acknowledging internally that wouldn't happen because it was "uneconomical."

The assessment was correct. And the contradictions between what the chemical giant told the world and what it whispered privately were a hint of the tensions at the heart of this new endeavor. Because while companies had begun to understand recycling's ability to ease people's concerns about plastic, the obstacles to doing it effectively are daunting. They're not just economic but also logistical and technical. Plastic must, after collection, be sorted by type. Even varying forms of a single kind—the PET in bottles and the PET in berry crates, for example—can't be reprocessed together. Different colors must be separated too. The bar for purity is high: Even a few missorted items can make an entire batch unusable. Today, there are thousands of plastics, and some that look the same are structurally distinct. More than ten thousand chemicals go into making them, so even similar types can contain additives that preclude their being combined. And many products and packages contain multiple plastics, often layered together or mixed with other materials like paper or adhesive.

That makes the necessary sorting so labor-intensive as to be all but impossible—and prohibitively expensive, especially since recycled plastic must compete with plentiful, inexpensive new material. "I mean, who's going to pull all these layers apart? Why would anybody pay to do that?" asked journalist Laura Sullivan, who investigated the industry's deception for NPR and PBS's *Frontline*.

There are big technical difficulties too. Unlike aluminum and cardboard, which can be recycled again and again, plastics' quality deteriorates with every go-round. So even the most recyclable types—PET drink bottles and milk jugs of high-density polyethylene—often become not new containers, but synthetic lumber, or textiles for carpets or clothing. Those materials cannot themselves be recycled, so really they just delay disposal. For anything other than bottles and jugs, the journey to oblivion is faster still.

And the benefits are smaller than we like to imagine. Turning an old can into a new one avoids the massive toxic pollution and energy use from the mining and smelting necessary to produce aluminum, just as recycling cardboard and paper saves trees. That makes those materials very much worth salvaging. But even so, recycling's own footprint is substantial. So it's no panacea even in the best of circumstances. Producing less packaging would always be better. For plastic, the equation is even starker. Recycling—which typically involves shredding sorted material, then melting it into pellets manufacturers can repurpose—is not only energy-intensive; there are also concerns it may release vast amounts of microplastics. And in poorer countries, where waste from rich ones often ends up, recycling workers, including children, can be exposed to dangerous chemicals.

The product that results is often hazardous too, even more so than virgin plastic. Recycling's heat creates new toxins, including cancer-causing dioxins. And discarded items carry contaminants they've come into contact with, including household chemicals like pesticides and solvents, into the new material. One review found recycled plastic bottles leached 150 different chemicals into drinks. A research group that bought recycled plastics in nearly two dozen countries reported none were free from harmful chemicals. Such hazards have been found in toys, kitchen utensils, and other common items made from recycled plastics.

THE DIFFICULTIES OF PLASTIC RECYCLING HAVE ONLY GROWN AS PACKAGing's volume and variety has exploded over the years. But industry understood the fundamental obstacles from the get-go. Speakers at the 1969 packaging waste conference we looked in on in chapter 2 summed up the central problem: "The aims of a package manufacturer and the conservationist not only fail to coincide, but conflict," packaging design executive Eric Outwater explained. Containers were easiest to salvage when they were composed simply, of one material, but packaging was becoming more complex. Another speaker offered a telling example: While a milk carton was mostly paper, its coating meant that "for a man who wants to re-use it in a paper mill, it might as well be made of plastics." Such combinations, packaging consultant Arsen Darnay said, "made packaging materials virtually unrecoverable after use." That was unlikely to change. Companies had started with simple packages, and "we found them wanting," so the future would bring "more and more exotic combinations of materials," he correctly predicted. "If reuse of materials" was a goal, "it is clear that trends in packaging technology shut out this avenue of approach."

Mobil gave its own clear-eyed summary in the early '70s of why less than 2 percent of municipal waste was being recycled: Sifting what was salvageable from what wasn't took work, "and when you get enough of it, you have to ship it to a plant where it can be scrubbed. Or purified. Or refined. Or upgraded. And then—maybe—you'll have a raw material almost as good as the nice, clean stuff a supplier can deliver to your factory door for a lot less money."

Those basic realities remained as recycling took off in the 1980s and '90s, and so did companies' understanding of them. The verdict from a 1986 industry analysis was clear: "Recycling is not and will never be commercially viable unless it is significantly subsidized by a government entity." It would be great if the plastics consumers tossed into recycling bins were turned into something usable, an Eastman Chemical representative said in 1994. But while "someday this may be a reality," he said, "it is more likely that we will wake up and realize that we are not going to recycle our way out of the solid waste issue." That warning anticipated almost word for word environmentalists' mantra today that we can't recycle our way out of the plastics mess. Even when plastic recycling was possible, the industry also knew, it didn't necessarily bring the benefits people expected. Often, one producer reported in 1993, "the energy and other resources consumed outweigh the environmental gain."

Occidental Chemical executive William Carroll offered up a handy scapegoat for recycled plastic's competitive disadvantage when he testified to Congress in 1992. It was companies' favorite one: the public, whom he blamed for failing to demand recycled material. "If you are a regular citizen," Carroll said, and "you haven't made the people who bring the products to you know that this is what you want, then you are not doing the whole job." Even so, he sneered at those who did express enthusiasm for recycling, saying they just wanted "penance for the sinfulness of being wasteful." And he dismissed as "draconian" the most effective way of overcoming the pricing problem—legal mandates that packaging include recycled material. "The creation of an artificial market based on ideology and not true demand" was "a deal with the devil," he said, likening such requirements to East German diktats about what kind of cars manufacturers could make. In its aggressive finger-pointing, snide condescension toward those concerned about the harm the industry was wreaking, and furious hostility toward any constraint that might be placed on them, Carroll's testimony perfectly encapsulated plastic makers' bad-faith messaging on recycling.

He summed up plastic recycling's perennial challenge neatly too. "The quality of virgin material was better—and it cost less." The industry also understood that the new plastic it was pumping out in ever-larger quantities would make it even harder for recycled to compete. "Virgin supplies will go up sharply in near future" and "kick the shit out of" recycled plastic prices, an American Plastics Council staffer scrawled during a 1995 meeting, in notes obtained by the Center for Climate Integrity. Others were frank about an even bigger obstacle. For plastic producers, a retired Procter & Gamble executive pointed out in 1996, recycling, if it could be made to work, would be competition. "They don't want to see it succeed."

But if its realities were underwhelming, recycling's power as a public relations tool was enticing. By the mid-1980s, industry could see both how much it needed such a tool and how appealing recycling was to the public. With officials around the country beginning to push restrictions on various single-use items, "the plastics industry was made to feel the pressure acutely," one executive later recalled. "The call was to recycle or be banned."

The Society of the Plastics Industry president Larry Thomas laid out the stakes in a confidential 1989 memo, *Frontline* reported. "The image of plastics among consumers is deteriorating at an alarmingly fast pace," he wrote. Polling showed "it has plummeted so far and so fast" that "we are approaching a 'point of no return'" beyond which "it will be impossible to recover our credibility." The proportion of people who saw plastics as harmful to health and the environment had jumped by fifteen points, to 72 percent, in just a year, Thomas warned. "Business is being lost; product growth rates are being dampened and stock analysts are beginning to take notice."

One way to remedy that would be a decade-long, quarter-of-a-billion-dollar ad blitz touting plastic's virtues. But it was also clear that "if the public thinks that recycling is working, then they're not going to be as concerned about the environment," Thomas, now retired, explained to NPR. A Mobil newsletter echoed that assessment: "As recycling programs prove they work, we expect demand for our products to grow." Industry lobbyist Roger Bernstein put it most pithily, explaining recycling was a "guilt eraser" that soothed consumers' ever-present anxiety about waste. "As soon as they recycle your product," he told author Susan Freinkel for her 2011 book, *Plastic: A Toxic Love Story*, "they feel better about it."

It wasn't just about easing consumers' worries but also stopping anti-plastic laws. "No doubt about it, legislation is the single most important reason why we are looking at recycling," the director of the Plastics Recycling Foundation, an alliance of dozens of big industry names, said in 1988. By that point, companies had begun urging states and cities to add plastic to their recycling programs. It might not have been industry's first choice, but if the alternative was bans of particular products or materials, recycling didn't seem so bad. Not only would it "divert public attention away from stronger reforms," author Heather Rogers wrote; it also "further normalized growing consumption, telegraphing to the public that even in the act of discarding one could be environmentally responsible." And by giving industry a tool that looked and sounded environmentally friendly, Rogers noted, it would unleash a new era of corporate greenwashing.

LEW FREEMAN WAS WORKING AT THE AMERICAN PETROLEUM INSTITUTE IN the late 1970s when a friend mentioned he'd be leaving his job running the Society of the Plastics Industry's Washington office, and asked if Freeman wanted to take over. He soon got hired and would spend more than twenty years at the society. On a call from the Appalachian home he's now retired to, he describes it to me as a loose confederation of companies across different parts of the industry. Back then, the society included the petrochemical firms making plastic, manufacturers turning it into goods and packaging, and suppliers of machinery and molds to those manufacturers. As time went on, he recalls, the petrochemical companies grew more active—and more powerful—within the group.

Freeman, who's warm and garrulous, in a plaid shirt and fleece vest, says he doesn't regret his time at the society, which later became the Plastics Industry Association. But since leaving in 2001, and watching the relentless spread of the material on whose behalf he once lobbied, he's traveled a very different path. During his first few years at the society, Freeman tells me, his lobbying focused mainly on concerns about plastics' flammability and restrictions on chlorofluorocarbons, chemicals that damage the atmospheric ozone layer and were used back then to make polystyrene foam, or Styrofoam. But as waste exploded onto front pages, petrochemical and packaging producers were getting worried. At a board meeting in early 1988, Freeman recalls, a DuPont vice president urged a more aggressive response to calls that something be done to reduce the flood of plastic trash. Before long, Freeman and his boss visited DuPont's Delaware headquarters, where, he tells me, one executive insisted, "If I had $5 million, I could solve this." The DuPont man didn't quite say it, but the implication Freeman heard was that such funds would go toward publicity, to "advertise our way out" of the problem. "That was the context."

For Freeman, it felt telling. Plastics companies, he believes, have been consistently unwilling to engage substantively with concerns about waste. From the beginning, their response was "more PR and hype and advertising than actually addressing the problem," or even acknowledging it. They "respond to these criticisms as if they were communications problems, informational problems."

Not long after the DuPont meeting, chemical executives created the Council for Solid Waste Solutions, under the society's auspices, to spearhead their response to the public clamor. Members included Exxon, Mobil, Chevron, Dow, and Amoco, and early on, Freeman says, there was little consensus on the best approach. "I'm not even sure there was consensus about what the problem was, other than they were being criticized in state legislatures and state regulatory agencies, and they didn't like that," he tells me. The council's goal, Freeman believes, was less to find solutions for waste than to "stem the tide of all of the anti-plastics" proposals. Recycling didn't emerge right away as a tool for doing that, but it would soon become a major focus, even though members expressed doubt about "whether it could work or not," Freeman recalls. "There was always, in my memory, skepticism and a lack of belief in the viability of plastics recycling."

Setting up organizations whose anodyne-sounding names gave no hint of their industry-dictated agendas was already a favorite tactic. A few years earlier, the society's members had funded the New York–based Plastics Recycling Foundation. The Council for Solid Waste Solutions itself would later become the American Plastics Council, and was eventually absorbed by the American Chemistry Council. The endless rebranding of entities that, for the most part, represented the same companies would remain a go-to, with industry-backed groups such as the Recycling Partnership, the Sustainable Packaging Coalition, and the Alliance to End Plastic Waste popping up in the years to come. (Neither the American Chemistry Council nor the Plastics Industry Association, the new name for the old Society of the Plastics Industry, replied to my emails seeking comment.)

Soon after its formation, the Council for Solid Waste Solutions created what one of its lobbyists called a "strike force" to help build recycling programs. Its point person headed to Minnesota, where Minneapolis and St. Paul had made national news by banning many kinds of plastic packaging. With hundreds of cities and organizations requesting copies of the 1989 ordinances, and the state legislature considering a similar bill, snuffing them out felt urgent. The council set up recycling centers, provided sorting machines, and created a "Blueprint for Plastics Recycling" to encourage local governments to add plastics to curbside collection programs. It seemed to work: The Twin Cities' bans would never be enforced.

One council staffer summarized the strategy vividly, writing, "We need to get out at the grass roots level & do guerilla warfare like our adversaries." Industry never intended to fund local programs for the long term. Companies' "attitude was, 'We'll set this up and get it going, but if the public wants it, they are going to have to pay for it,'" explained the executive in charge.

Many of the recycling plants big plastic makers began building in the late '80s fell squarely into a category the Center for Climate Integrity, a research and advocacy group, calls "performative investments"—that is, they achieved more in publicity terms than practical ones. One of the highest-profile efforts was the National Polystyrene Recycling Company, a joint venture of Chevron, Mobil, Amoco, and other makers of polystyrene, or Styrofoam. Eager to rid the material of its environmentally villainous reputation, the new company vowed to recycle 110 million kilograms of the stuff annually, encouraged municipalities to collect it, and was soon getting foam plates and cups from school cafeterias. Polystyrene makers trumpeted the work in ads, but within five years the recycling company began shutting plants and laying off staff; it was sold after five more, its president blaming the public, which, he said, "does not want to buy recycled products." The new owner soon closed its last remaining facilities.

But industry had learned something valuable: It was easy to make a big splash with an ambitious numerical target. In 1991, the Council for Solid Waste Solutions set its own goal. By 1995, it announced, it wanted a quarter of the country's discarded plastic bottles and rigid containers recycled. Some thought 10 percent would be more realistic, but the group decided 25 percent was "the lowest rate that would be acceptable to the general public," one executive later recalled.

For all the fanfare, Lew Freeman tells me, industry's efforts fell far short of what would have been needed to make plastic recycling succeed—if that was even possible. "They never developed the kind of infrastructure" that the glass or aluminum industries had nurtured for recycling those materials, he says. "There was not a dedication to investing" in it the way those sectors had done. Of course, plastic recycling faced obstacles other materials didn't. But even so, a serious, well-intentioned push might have nudged rates higher, at least for bottles and milk jugs, the most recyclable items. In Freeman's view, companies were always more interested in talking up the money they were spending than in being honest about the mostly anemic outcomes. An Exxon representative summed it up best: "We are committed to the activities, but not committed to the results."

As for Freeman, after he left his job at the plastics society, he and his wife eventually moved full-time to the home that had been their weekend getaway, in a tiny northwestern Virginia town. With fracking taking off across Appalachia, he founded an environmental alliance to campaign against a pipeline that threatened to despoil his beloved Allegheny and Blue Ridge Mountains. He won that fight—the pipeline was never built. It was a remarkable turnabout for a man who'd worked early on at the American Petroleum Institute. Freeman spent far longer with the plastics society, and he couches his criticisms of that industry in gentle terms. His critique is persistent, though—and public. It's been disappointing, he tells me, to watch the companies he'd worked with refuse to engage seriously with legitimate criticism, even as the volume of plastic they churn out has climbed to levels almost unimaginable in the 1970s. Listening to companies repeat the same old talking points all these years later, he says matter-of-factly, "I kind of shake my head."

WHATEVER SERIOUSNESS OF PURPOSE MAY HAVE BEEN LACKING IN COMPANIES' efforts to stand up lasting recycling systems, there was no shortage of enthusiasm for boasting about them. Industry knew it needed to, as one executive said, "provide visibility" for recycling. So it unleashed a torrent of advertising to tout its promise and gloss over the difficulties.

Just a few months after *Time* put an image of what it called the "Endangered Earth," wrapped in plastic, on a 1989 cover, the Council for Solid Waste Solutions bought a million-dollar, twelve-page advertorial titled "The Urgent Need to Recycle" in the magazine. "Once is not enough," it exhorted, offering an idyllic description of a child running on a recycled plastic dock. Soon, it gushed, landfilling would be "a last resort." Companies repeated such promises endlessly. One Mobil ad

"GUILT ERASER"

promised that soon, "even fewer plastic grocery sacks will wind up as garbage. Instead, they'll be recycled into new, useful plastic products." Anyway, it added, plastic waste wasn't just industry's fault. "Every American throws stuff away, and every American is therefore part of the country's nagging solid waste problem. Mobil Chemical, we're proud to say, is also part of the solution." That wasn't the only time a sour note snuck in. Another of the oil and chemical giant's ads complained those working in plastics were treated like "environmental villains" and told "there ought to be a law against the things you make," because of a wrongheaded belief they couldn't be recycled. Rejecting that view, it said industry wanted half of PET bottles recycled within three years.

Other ads similarly sought to bat down suggestions plastic might not be recyclable. One in *State Legislatures*, a magazine for lawmakers and their aides, cast those "who would have you believe that the sky is falling on plastics recycling" as Chicken Littles jumping to an "erroneous conclusion." A 1993 magazine ad exulted, beside a picture of a refrigerator crowded with plastic packaging, "Your new carpeting may already be in your refrigerator." Plastic bottles, it continued, were "turning into toys, pillows, garbage cans, sailboat sails, even plastic 'lumber.' Not to mention back into new bottles." Styrofoam cups were becoming "building insulation, office accessories and VCR tape cassettes." Those claims prompted eleven state attorneys general to sue, claiming the $18 million campaign misled readers. In a settlement, the American Plastics Council paid $110,000—barely pocket change for a vast industry—and promised to caveat future ads by warning recycling facilities might not be available everywhere.

Even that modicum of accountability was rare. For the most part, the blizzard of upbeat claims about plastics' recyclability went unchallenged. One avenue for making them ran through classrooms. Industry created resource packs, films, and activity booklets it provided to schools for free or a nominal cost. Videos like the Plastic Bag Association's

PLASTIC INC.

"Don't Let a Good Thing Go to Waste" and Keep America Beautiful's "Mister Rogers' Recycling Video" brought children the same message TV and magazine ads were feeding their parents. "Most plastics can be melted and reused over and over again," the American Plastics Council fibbed in one film. Another, for middle- and high-schoolers, would "inspire students to get involved in recycling," said the council, which also created a how-to guide for setting up a school recycling program. A council spokeswoman acknowledged the obvious in discussing one video: "It is propaganda."

It seemed to work. By the mid-1990s, recycling had been "elevated to the status of motherhood and apple pie," historian Susan Strasser wrote. It looked to many like the best remedy for the ills of modern consumerism—the end-all and be-all of environmental protection. Environmentalists were on the bandwagon too. Recycling, it was clear, made people feel good. "A means to get rid of things with a clean conscience," in Strasser's words, it was a small action that gave us the sense we were doing our part for the planet.

That sentiment—one industry had worked so hard to sell—exerted a powerful force. "The magnetic, gravitational power of recycling," one Oregon environment official later said, led "policymakers and the public to just talk more and more" about it, "and less and less and less about anything else"—notably, any suggestion of reducing the amount of plastic being made. And those policymakers backed their words with action. With half of states passing recycling laws and the number of localities offering curbside collection skyrocketing, the late 1980s and early '90s were the stretch when, as one historian put it, "this thing just becomes everywhere."

Indeed, by the late '90s, polling found the public, government officials, and the media all saw recycling as the best approach to handling plastics. Tellingly, opinion was shifting the other way among waste management experts, who were losing faith in its viability. But all the groups polled believed plastic could be recycled at much higher rates than were realistic. The good feeling seemed to be reflected onto the material itself—people saw the industry much more positively than they had a few years earlier.

Companies' achievement was twofold. Not only had they persuaded us to stop questioning why throwaway junk had proliferated so widely. We'd also accepted that recycling it all should be our responsibility, not theirs. It was a weighty burden. At the turn of the millennium, US recycling programs' revenues covered only about a third of their costs, with the rest shouldered by tax- and ratepayers. The public was handing an enormous—if largely invisible—subsidy to beverage giants that paid nothing for the waste they created, Coke chronicler Bartow Elmore argues. While industry would continue to tout its support in the years to come, he wrote, "corporate recycling programs have been built on infrastructure that it took municipalities decades to construct."

FOR INDUSTRY, THE RECYCLING BLITZ'S SUCCESS BROUGHT NOT JUST GOOD feeling but tangible results: "The environmental pressure is off," a plastics consultant crowed in 1996. Lobbyist Roger Bernstein agreed. "There's a shift in the political climate," he said, with plastic no longer "singled out" for vitriol. The "anti-packaging forces stirred up by 'environmental hooligans' were now in retreat," another industry figure cheered.

With their political objectives met, the makers and users of plastic were free to drop the goals they'd unveiled a few years earlier. The American Plastics Council backed away from its 25 percent recycling target, its spokeswoman dismissing "the idea of rates, dates, mandates" as "very artificial." Such numbers had been helpful communications tools, the council's president elaborated, but industry "has progressed beyond" them. Companies understood the risk in abandoning a goal they'd shouted so enthusiastically. Failure, a council staffer noted while meeting with an Exxon executive, would be "HIGHLY SENSITIVE POLITICALLY." But even unmet, the target had served its purpose. And like the creation of green-sounding front groups, the setting of such lofty goals would become a favorite industry tactic, with new ones proffered when political pressure heated up, only to be forgotten when it died down.

As the council's target fell by the wayside, recycling facilities closed. Within two years of unveiling a joint venture to process used bottles, DuPont and the trash giant Waste Management had both pulled out. They'd been paying $1,500 a ton to collect and process material with a market value of eighty to one hundred dollars. "We could not see our way clear to a reasonable financial return," a DuPont executive said. A bottle processing facility in the Bay Area survived barely a year, and an Oregon recycling company sold for scrap a $1.5 million sorting machine it had gotten from the Plastics Council. One investigation found nearly a dozen projects that opened in 1989 and beyond—including headline-friendly programs to recycle plastic from schools and national parks—had failed within a few years. Others had been announced but never built. Overall, *Plastics News* observed, "recycling appears to be jinxed."

One question continued to nag. Knowing recycling might cannibalize sales of new plastic, did the industry even want it to succeed? Larry Thomas, the former Society of the Plastics Industry president who spoke to *Frontline*, thought not. "Nobody that is producing a virgin product wants something to come along that is going to replace it," he said. A Milwaukee plastic recycler was more vehement, slamming the American Plastics Council as "antirecycling wolves in sheep's clothing." The "people who set those goals," he said, "were the least desirous of seeing those goals met."

Whatever companies' intentions, one move draws particular scrutiny from critics who contend industry actively sought to sabotage recycling. In 1988, the Society of the Plastics Industry created a labeling system assigning numbers to different kinds of plastic: 1 and 2 for PET bottles and the high-density polyethylene typically used in milk jugs, 3 through 6 for other, harder-to-recycle types, and 7 as a catch-all for everything else. Known as the Resin Identification Code, it was ostensibly intended to help recyclers sort plastics. The numbers would be placed in the middle of a triangle formed by three chasing arrows—a flattened, simplified version of the famous recycling symbol Gary Anderson had designed decades earlier.

There were concerns from the start. As one industry group's fact sheet acknowledged, labels would be "of limited practicality" in facilitating sorting because most packaging contained multiple types of plastic. Even if effective sorting was possible, the resin code's categories were too broad to be useful. The biggest worry was the triangle's resemblance to the universally recognized recycling symbol. Its presence would clearly suggest items "are made of recycled material or that they are recyclable," Connecticut's environment department wrote, although that was "generally not the case." Consumers would treat anything labeled with the chasing arrows as recyclable, and flood local systems with material they were unequipped to handle, the department warned. "This will have a severe impact on the already marginal economic feasibility of recycling plastics" and on recycling programs as a whole.

Brushing aside such worries, the Society of the Plastics Industry pressed states to require its code's use, and thirty-nine soon did. As one industry newsletter explained, legislatures adopting the coding system sometimes did so "as an alternative to more stringent legislation," like bans on disposable items. A California bill originally written to mandate that plastic packaging be recyclable or biodegradable, for example, was changed to simply require the code be printed on it.

The symbol quickly began appearing on just about every plastic container, and the problems were apparent immediately. Suddenly, people had begun putting all kinds of plastic, not just bottles, in their bins,

San Diego recycling center operator Coy Smith told PBS. They'd point to the symbol, insisting, "It says it's recyclable," he recalled. He knew that wasn't true, that in reality, "there's no one that would even take it if I paid them." The story was the same everywhere. And it wasn't just that facilities couldn't recycle the new plastic coming in. The deluge made it impossible even to process material that could have been recycled if well sorted.

Smith met with industry representatives and, along with other recyclers, sent them a report describing the problems. The code, it said, was creating "unrealistic expectations." Smith and others urged the society to drop the triangle to avoid confusion with the original recycling symbol, but it didn't happen. To him, it was clear why. The perception of recyclability made products more attractive to consumers. Companies, his report had warned, were using the code "as a 'green' marketing tool,'" to sell more plastic.

THE WIDESPREAD BELIEF THAT JUST ABOUT ANY KIND OF PLASTIC COULD, and should, be recycled was soon cemented. Martin Bourque watched it happen—and then saw the impact. A trim man with a neat, graying beard and curly salt-and-pepper hair brushed to one side, Bourque runs the Ecology Center, a nonprofit group that operates Berkeley, California's recycling collection. I meet him in an open, sunny office, upstairs from a shop selling a wider range of reusable, refillable, and minimally or alternatively packaged products—from teeth-cleaning tablets to sunscreen sticks—than I've encountered before.

As the millennium turned, and wore on, much of what Americans and those in other wealthy countries tossed in their recycling bins was ending up in China. Its factories could make use of the most recyclable plastics, turning them into clothing, carpets, planks, bins, and more. Back then, China also had low-wage workers ready to sift such material from the mountains of unusable plastic manufacturers didn't want. Shipping was cheap too. The container vessels delivering Chinese-made sneakers, TVs, car parts, and toys to the world were going back empty, so it didn't cost much to load them with waste. Those forces, Bourque tells me, combined to create an "amazing sucking action" pulling not just plastic scrap but also paper, cardboard, glass, and aluminum to China. Those tasked with getting rich nations' waste recycled—local governments and the companies contracting with them—were happy to sell what they collected to export brokers. It was easy not to ask what happened to it.

China's plastic scrap imports climbed from near zero in the late 1990s to 110 million metric tons a year by 2015. The United States alone was sending 1,500 shipping containers of waste there every day. Europe was exporting half its plastic recycling, with 85 percent of that going to China. The network that sprang up to process it was vast, composed largely of informal, unregulated facilities. In Wen'an, outside Beijing, twenty thousand plastics processors clustered in just a few villages, where workers sorted, shredded, and melted down usable material with rudimentary equipment and little or no protective gear. Unwanted leftovers were simply dumped or burned. Polluted wastewater ran into streams, killing fish, and destroyed crops when used for irrigation. Residents suffered from lung ailments and other illnesses.

Bourque knew much of what was getting shipped overseas was worthless. But encouraged by companies' relentless messaging to believe the opposite, people wanted all their plastic waste recycled—the pouches, tubs, tubes, clamshell containers, cups, wrappers, and more. And in the early 2010s, he says, industry was sending representatives to recycling conferences to urge localities and their waste haulers to collect it. The message landed.

One by one, neighboring cities added more plastics to their collections. The Ecology Center was already picking up Berkeley's bottles and milk containers, but pressure to take the rest was growing, from "residents saying, 'Hey, we want to recycle this stuff,'" he recalls. The local press "called us troglodytes" for refusing, and when the group finally gave in, wrote "'Ecology Center comes out of the caves'—literally the headline said that," Bourque says with a laugh. The Ecology Center had relented only after finding a Guangdong province facility with decent worker and environmental protections. But the arrangement didn't last. In 2013, China began tightening enforcement of its rules on scrap imports' quality, and the company taking Berkeley's plastic shut down. The crackdown was roiling waste markets globally. Within a year, the price that mixed plastics could fetch dropped from thirty-five dollars a ton to zero. Soon, cities had to pay to get rid of it, yet another burden for local governments and ratepayers. "Those folks were getting so screwed financially," one state senator told me. The Ecology Center began working with a broker who sent its waste on to another broker, a common arrangement that meant "you have no frickin' idea where it goes," Bourque says.

In 2014, the documentary *Plastic China* debuted online, casting a painful light on what was happening to wealthy countries' recycling. A longer version, released later, focused on two Shandong province families whose members lived beside piled-up foreign plastic they sorted in grim conditions. It shows children and adults stuffing scrap into shredding machines, melting it into thick goo, and breathing air dense with black smoke and bits of plastic. In one scene, a woman gives birth amid the piles of waste.

The film shocked many Chinese. As the country grew wealthier, it was less willing to abide informal recycling's awful pollution and dangerous working conditions. Also, labor wasn't so cheap anymore, and China's own consumers were using more plastic, so factories could source the recycled material they wanted domestically. China, by then the world's largest recycler, announced it would close its doors to plastic scrap in 2018.

The policy was called National Sword, and it upended plastics recycling around the world, laying bare the shaky economics, willful blindness, and fundamental dishonesty it was built on. From Australia and Japan to Britain, Germany, and the United States, discarded plastic piled up. Waste tossed in recycling bins had to be landfilled or incinerated. Some local governments scaled back recycling programs or shuttered them entirely. "It was a huge wakeup call," Bourque says. The realities of plastic recycling were becoming impossible to ignore.

THOSE REALITIES WERE A FAR CRY FROM WHAT INDUSTRY HAD PROMISED. For all its happy talk, only 9 percent of the world's plastic waste has ever been recycled. With more produced every year, even a flatlining of that rate means the amount getting landfilled, incinerated, or littered keeps climbing. But as China tightened its rules, the rate fell in much of the world. Even at its height, US plastic waste recycling had never broken 10 percent, and an analysis a few years after National Sword put it below 6 percent. Rates are better elsewhere, but not by a lot. Australia, for example, reports recycling about 14 percent of plastic waste. In any case, such statistics typically reflect what is collected for recycling, not what is actually recycled. Since poor sorting or other contamination forces processing plants to reject a great deal of plastic, the true figures, nearly everywhere, are usually even worse than the low levels reported officially.

The biggest bright spot is PET bottles. Britain, for example, claims to recycle nearly three-quarters of them (although an independent analysis found government figures overstated rates for other plastics by a third). The rest of Europe achieves similar levels, and is aiming even higher. In the US, the rate has typically been just below 30 percent—not amazing, but not nothing. The ten states that mandate bottle deposits account for a disproportionate share, a reminder that the bottle bills industry opposes so vociferously are the best way of retrieving clean, well-sorted plastic.

Companies' answer to the dispiriting overall reality has been to aggressively promote a set of technologies known as chemical recycling (in contrast to the traditional shredding and melting, which is called mechanical recycling). Industry prefers the name "advanced recycling," but in reality chemical recycling techniques have mostly been around for decades. The predominant method is called pyrolysis—essentially superheating plastics in an oxygen-free chamber to break them into gases and liquids including naphtha, which can be used to make new plastics. In companies' rosy vision, this paring back of discarded plastic to its constituent molecules enables those molecules to be recombined into high-quality new material, in a process that can be repeated endlessly.

With a huge advertising push reminiscent of those that have inflated hopes before, companies are selling such techniques as a way to give new life to previously hard-to-recycle plastics—and a solution, at long last, to the world's plastic problems. "It used to be that only SOME plastics could be recycled," ExxonMobil said in a social media post typical of that framing. But with "our revolutionary advanced recycling technology," it boasted, "we can give any plastic a new life. A chip bag into a Ferrari? A water bottle into a computer? Is it Science Fiction? It's happening now." Unlike with mechanical recycling, one executive wrote, "there are no evident technical limitations regarding how many times a plastic product can be put through advanced recycling processes."

The glossy promises, unfortunately, don't survive contact with reality. Very little of what chemical recycling plants put out—15 to 25 percent at best—becomes new plastic. At the plant ExxonMobil's gushy tweet referenced, it's just 8 percent. The rest is fuel, such as diesel, and other chemicals. And even most of what goes into the process isn't plastic. Because it's contaminated by packaging's many additives, naphtha from waste can be used to make new plastic only when combined into a mixture that's 90 percent oil-derived naphtha. So this form of recycling is just fractionally different from virgin plastic production. The plastics that result contain at most 10 percent recycled material; in practice, that can be less than 5 percent. One government study found plastics produced through pyrolysis and a related technology called gasification retain as little as 1 percent of the waste plastic put in. New plastic from ExxonMobil's plant, near Houston, contains at most .09 percent plastic waste—less than a tenth of a percent, or one part in a thousand, according to company documents obtained by California's attorney general.

None of that is grist for effective marketing. But industry has a neat trick for puffing up chemical recycling's achievements. Known as "mass balance accounting," it shifts—on paper—all the recycled plastic a plant produces into just a portion of the new material. By not making recycling claims for some of the plastics they sell, companies can claim a higher proportion of recycled content for the rest. One analysis called the approach "mathematical acrobatics." Such flips and cartwheels mean goods advertised as "made with 50 percent recycled material" may contain far less. With a twist called "free allocation," companies can even count the waste that became fuel as "recycled content" in new plastic, entering, as California alleges in a lawsuit against ExxonMobil, "the imaginary realm." (The company sued the state's attorney general back, accusing him of "blatant misstatements and attacks on Exxon-Mobil's character" and defending chemical recycling as "a proven technology" that can keep plastic out of landfills. ExxonMobil's spokespeople did not reply to my messages seeking their comment for any aspect of this book.)

Without mass balance accounting, chemical recycling would collapse under its own weight. Prohibiting it, an executive at the petrochemical giant INEOS told me, would be tantamount to "putting a nail in the coffin" of the process's viability. That's why arcane-seeming methodological questions have become a fierce political battleground, with industry pushing lax rules that would help it use chemical recycling to meet requirements countries and states are increasingly imposing for some new plastics to contain recycled material. In Brussels, a European Parliament aide said the accounting debate had become "a whole war" happening "behind closed doors." Big plastic makers like ExxonMobil and INEOS are among those already using an unregulated, voluntary system authorizing "certified circular polymers" labels for plastic that may contain little recycled material. For consumer-facing companies like Nestlé, L'Oréal, and Procter & Gamble, chemical recycling offers a way to meet public pledges to use more recycled plastic.

But chemical recycling not only fails to live up to its hype—the process itself does a great deal of harm, generating even more carbon emissions than making plastic from oil and gas. Industry often claims the opposite, but plastic from chemical recycling scores better only if you assume the virgin plastic ends up incinerated. Even so, both American and European officials are underwriting chemical recycling plants with millions in subsidies.

For all its failings, chemical recycling offers companies what one activist aptly called "a narrative solution"—or more pointedly, a comforting story deployed to ease our qualms. As ever, industry knows it needs an answer to consumers' nagging waste concerns, and as ExxonMobil internal communications obtained by the California investigation noted, the public seems "favorably receptive to advanced recycling messages." Such messages could, the company believed, help "avoid the 'negative' impacts/consequences" of "the circular economy way of thinking." That thinking, it noted, could become a "market threat" for its chemical arm—80 percent of whose growth depends on single-use plastics. Even if chemical recycling efforts turn out "to not be financially sustainable," the company concluded, "the public perception benefits received will be invaluable."

FUNDAMENTALLY, PLASTICS COMPANIES' MESSAGES HAVEN'T CHANGED much over the years—although they're using new avenues to get them

"GUILT ERASER"

out. These days, industry is paying TikTok and Instagram influencers to talk up recycling. One chatted in a sponsored video about bringing water to keep her dogs hydrated on a picnic, explaining, "I always opt for PET plastics because they're 100 percent recyclable." The industry group behind such posts, NAPCOR, represents dozens of big companies, including INEOS, the major packaging producer Amcor, and Dart, which makes the distinctive red Solo cups. The online campaign sprang from companies' desire to push back against a "tide of anti-plastic sentiment," one internal document said, echoing language industry's been using since the 1970s. Another note, about the 2024 Paris Olympics' prohibition of single-use plastic bottles and cups, also could have been written any time in the past five decades: "Recycling, not banning," it read, "is the best way to eliminate plastic waste."

Despite the never-ending barrage of misinformation, recycling has a valuable role to play. For other materials—aluminum, cardboard, paper, glass—it's both viable and important. Even for plastic, especially those most-recyclable bottles, it can be useful, if only in a least-worst sense. And there are ways to improve it: Deposits to get containers back, laws pushing industry to simplify packages and pay for recycling, mandates to use recycled material in new packaging. And, crucially, honesty about its limitations. In California, state Senator Ben Allen sponsored a law prohibiting the chasing-arrows logo on any item for which recycling was available in less than 60 percent of the state. Companies shouldn't "hide behind that symbol just because their product is theoretically recyclable at a thousand degrees at some lab," he told me. "Unless it's really recyclable under real-world conditions," a package shouldn't carry the three arrows.

Another key is that recycling can't stand on its own. To be meaningful, it must be accompanied by a reduction in the amount of plastic made and sold. Because it turns out industry was right about one thing: Recycling can be a potent eraser of our discomfort with tossing stuff away. One study found people used twice as much paper when a recycling bin was nearby, its presence seeming "to act like a subconscious green light to consume more," the author Franklin-Wallis writes. While one version of a circular economy is "about slower consumption, about valuing things, about consuming less and polluting less," he observes, "there is also another, more corporate vision, in which circularity simply encourages disposability." Indeed, as I wrote this chapter, Coca-Cola—its past littered with unmet recycling targets—abandoned plans to package a quarter of drinks in reusable containers, in favor of yet more recycling promises.

Martin Bourque reminds me of something that should be obvious. Recycling only matters if it averts the use of virgin resources—protecting trees from being turned into paper and cardboard, keeping oil and gas in the ground. If it simply underpins more consumption, "then it's being used for greater harm than good."

CHAPTER 5

"A Fantastic Window of Opportunity"

1976—The many varieties of plastics become, cumulatively, the most commonly used material in the world. Fifty-nine million metric tons of plastic produced.

In 1930, the Standard Oil Company of California was hungry for Middle Eastern oil. The British dominated drilling in Iran, and several other American companies had bought into Iraq, so Standard Oil of California, known as Socal, was prospecting on Bahrain, a small Persian Gulf island. The company sent Minnesotan geologist Fred Davies to lead its hunt, and his team soon struck oil beneath an oval-shaped hill. Two years later, Davies gazed across the water at a similar-looking mound on the Arabian peninsula—where the Saud family was just consolidating control, naming its territory Saudi Arabia—and wondered if petroleum riches might lie buried there too.

British engineers had scouted unsuccessfully beneath the peninsula's vast deserts, and even the Sauds didn't think there was anything there. But the royal family needed cash, so King Abdul Aziz was happy to let the Americans look. Socal agreed to pay £55,000 for that permission, plus another £100,000—along with royalties—if they discovered oil.

"Another corner of the world has been opened to American commercial interests," *The New York Times* proclaimed. By 1938, after repeated failure, company executives were giving up, but the local crew decided to dig well number 7 a little deeper. The decision paid off. They found a big, rich oil field, then another nearby.

The Saudi crude was high in sulfur, and none of the refineries Socal normally used could process it, so executives brought another US company, Texaco, into their venture. After World War II, both firms were short on capital, and they invited two more companies to buy in—Standard Oil of New Jersey and Standard Oil of New York. (The Supreme Court had split John D. Rockefeller's Standard Oil monopoly into more than thirty companies. Socal would eventually become Chevron, while the New Jersey and New York entities evolved into Exxon and Mobil, then ExxonMobil). The four US corporations named their partnership the Arabian American Oil Company—Aramco, for short.

Over the years, they found new reserves, and as operations expanded, the company and the Saudi government built pipelines, deep-sea ports, and a miniature American city at Dharan. The economy, and the national treasury, grew along with the drilling. But the royal family wanted more, and by 1950 Saudi officials were pushing to renegotiate their deal with Aramco. The company resisted, but events nearby in Iran—where lawmakers wrested control of their country's oil from the British—soon focused executives' minds.

For Aramco, such forcible nationalization was the nightmare scenario. (Western governments found the seizure deeply unnerving too; it helped prompt the CIA and British intelligence to orchestrate the overthrow of Iranian Prime Minister Mohammad Mosaddegh.) The Saud family—knowing they'd need American help to keep oil flowing until they could train their own workforce—wanted to avoid it too, author Ellen Wald writes in *Saudi, Inc.* Even so, the specter of an Iran-style grab helped Saudi Arabia press Aramco into a fifty-fifty profit-sharing arrangement, the best any Middle Eastern country had yet secured. It worked out for the Americans too—the oilmen, if not taxpayers back home. Aramco structured its payments to the Sauds as a tax, which it could then subtract from what it owed the IRS. That, Wald writes, meant the company simply "traded its American tax bill for a Saudi one," so sharing profits essentially cost it nothing.

The Sauds weren't done. What they really wanted was not just profit but control of their own oil. After years of pressure, the first two Saudis joined Aramco's board in 1959, and in 1972, the company agreed Saudi Arabia could buy in—starting with a 25 percent share, for a price believed to be about $500 million. By 1980, the Saudi government had purchased the whole company, at a cost estimated around $2 billion. This was a business deal, nothing like the confiscation Aramco once feared. The United States would still have access to Saudi oil, along with a strong incentive (despite its ally's dismal human rights record) to maintain close political and security ties—an arrangement that would shape US-Saudi relations for decades. The royal family—which knew having money to spend on schools, roads, and other necessities was critical to keeping its grip on the throne—was happy to overlook differences with its powerful friend too. Indeed, the Sauds had always prioritized business over the Middle East's tumultuous politics, and their ties with America had weathered the tensions of the 1973 Arab-Israeli War. The country would navigate the 1990s Gulf War too, and even the attacks of September 11, 2001, carried out by Saudi hijackers.

Only in 1988 did Aramco drop its Delaware incorporation and become a fully Saudi entity, renamed the Saudi Arabian Oil Company—known more commonly as Saudi Aramco. That same year, the first Saudi CEO replaced the last American one. Ali al-Naimi was a pearl diver's son, born three years before the company's first oil strike. He'd tended sheep until a half brother brought him to a company-run school, where he learned to read Arabic and English. Aramco later sent him to Lebanon, then the United States, for further study, and he returned to work his way up the corporate ladder.

When al-Naimi took over, Saudi Aramco was already the world's biggest oil producer. He pushed to expand globally, buying into refineries in the United States, South Korea, the Philippines, and Japan. He wanted not just growth but diversification: The refineries were a step toward shifting the business from just pumping oil to processing it too. And since they each agreed to buy Saudi crude, the purchases would also "guarantee a market," al-Naimi explained.

In 2022, when tech stocks swooned and oil prices spiked, Saudi Aramco surpassed Apple to become the world's most valuable company. That didn't last, but it remains, at the time of writing, the third-biggest by revenue, raking in nearly $500 billion a year. It's valued at nearly $1.7 trillion—almost four times as much as ExxonMobil, the next-largest oil company. Those riches have made Aramco "the engine that keeps Saudi Arabia running," Rice University energy expert Jim Krane told me. About 40 percent of the country's GDP and two-thirds of government revenue come from oil. The royals know that stream of cash, and the prosperity it buys for ordinary Saudis, is what keeps them in power. "Protecting the business," Krane wrote, is "a strategic and economic imperative for the kingdom and its ruling family."

That's why—despite its own vulnerability to extreme heat—Saudi Arabia sees the prospect of global climate action as, in the words of one assessment, "more of a threat than climate change itself." Publicly, Crown Prince Mohammed bin Salman, the country's de facto ruler, says the world "needs to go much further and faster in combating climate change." But together, the government and its main oil producer have played an outsize role in bringing us to the brink of planetary catastrophe. An analysis of history's biggest cumulative greenhouse gas emitters found Saudi Aramco was the number one corporate source. The only two entities ranked higher than this gigantic company were countries:

"A FANTASTIC WINDOW OF OPPORTUNITY"

China and the former Soviet Union. Industry disputes the methodology of such analyses, saying oil and gas companies are responsible only for emissions created by extracting and transporting their product, and not for the carbon dioxide that results when customers use those fuels. "We only produce oil because people want to buy it," an Aramco vice president said.

Indeed, Saudi Aramco—which *The Guardian* once called "the most profitable and the most polluting company of all time"—is a master of sustainability sleight of hand. "Environmental stewardship has always been part of our modus operandi," one executive said. The company talks up its use of drones rather than trucks when prospecting for oil and highlights work restoring coastal mangroves even as it vows to keep selling oil "for generations to come." Even Saudi Arabia's much-ballyhooed plans to get more of its own power from solar are mainly a way to free up oil for export.

All the while, behind the scenes, both country and company are working relentlessly to stop, or at least slow, any move away from fossil fuels. They're a forceful presence at global climate negotiations, where the Saudi delegation—dominated by the Ministry of Energy, which is deeply intertwined with Aramco—has spent decades throwing procedural wrenches into the works. A "thirty-year record of obstruction and delay," one report called it, aimed at ensuring that "U.N. climate talks achieve as little as possible, as slowly as possible." Saudis kill time with long speeches, object to sessions running over or happening simultaneously, argue against discussion of any topic not on the agenda. Sometimes, they "simply refuse to engage," the report said. "They will block an issue and then leave the negotiating room, paralyzing discussions." That work is made easier by a prize the country won at the very first climate summit, in 1992: a requirement that decisions be reached not by a majority of the nearly two hundred participating countries but by consensus—which is not quite unanimity, but close.

Other nations at the annual sessions have opinions on how fast or far a shift off oil, gas, and coal might go, one former UN adviser told *The New York Times*. But Saudi Arabia "doesn't even want to have the conversation." Indeed, its energy minister once called the very idea "a sequel to 'La La Land.'" The reason is clear: "The more they postpone, the more they earn," one researcher wrote. For the most part, the country has backed off its long-standing campaign to cast doubt on climate science, instead embracing a more polished stance of paying lip service to the problem's existence but promoting fossil fuel–friendly solutions—accepting the diagnosis while eschewing the most effective treatment. Old habits die hard, though, and as recently as 2022 Saudi delegates demanded deletion of the phrase "human-induced climate change" from a UN document. While there's plenty of blame to go around, Saudi Arabia's determined opposition is a big part of why the global negotiations have been such a fruitless grind, their snail's pace lagging far behind the rapid spiraling of climate disasters.

Saudi Arabia's effort to "keep the world hooked on oil," as one headline put it, stretches beyond its undermining of climate talks. It directed $2.5 billion into American universities over a decade, funding, among other work, research aimed at discrediting electric vehicles and boosting gasoline-powered ones. Across Africa, the Saudis are pouring money into investments aimed at ensuring new cars, buses, and ships run on fossil fuels, not electricity, a push one official promised would deliver "an oil uplift for the kingdom," the Centre for Climate Reporting found.

But even as it works to postpone the day the world burns less oil, Saudi Arabia is also preparing for it. Relying so heavily on a notoriously volatile commodity was always risky, and now, with the possibility of demand eventually cratering, it feels even scarier. In 2016, the crown prince announced Vision 2030, a plan to reduce oil dependence by diversifying and modernizing the country's economy. Aramco would be a key player in the effort. Perhaps Saudi Arabia's most plentiful resource could still fetch a profit—maybe an even greater one—in a new form. That, of course, is where plastic comes in.

IN 2017, SAUDI ARAMCO, IN PARTNERSHIP WITH DOW CHEMICAL, FINISHED WORK on a $20 billion, twenty-six-plant petrochemical complex called Sadara. The largest such facility ever built in a single phase, it sits on six square kilometers beside the blue-green Perisan Gulf, made from almost 30,000 cubic meters of concrete, as much steel as two Golden Gate Bridges, and 2,500 kilometers of pipes. Capable of producing more than three million tons of plastics and other chemicals annually, the complex is "one for the record books," Saudi Aramco boasts, yet another demonstration of its ability to execute "extreme engineering projects."

It wasn't the company's first step into plastics—it operates other plants, at home and abroad, with partners such as ExxonMobil, France's Total, and China's Sinopec. But Sadara "is not just another project," Saudi Aramco explains on its website. Instead, it is "the cornerstone" of plans to transform itself from an oil company into an "integrated energy and chemicals enterprise" that will be "the global leader in petrochemicals production." Doing so, said CEO Amin Nasser, would enable it to "derive the maximum value from every hydrocarbon molecule." Or, in plainer English, to wring as much cash as possible from the kingdom's buried riches. Saudi Aramco envisioned another complex nearby—PlasChem Park, where manufacturers could process the ingredients Sadara and other petrochemical complexes produced into finished products, from tires and auto interiors to diapers, flooring, packaging, and appliances.

It was, the *Financial Times* observed, "a powerful statement of intent." The company was heeding the crown prince's words. Plastic and other petrochemicals would strengthen the nation's economy by diversifying it, guaranteeing a healthy new stream of revenue, creating jobs in the plants themselves and hopefully attracting more in factories like

PlasChem Park's. Leaning into plastic would not only prepare one of the world's top oil powers for life after oil but also enable it to add value to its own crude, rather than leaving the job to others. In that way, this new push was really just another step in Saudi Arabia's decades-long project of claiming control of its own resources.

There was only one problem. This effort to move away from oil was built on oil—"attached by an umbilical cord" to Saudi reserves, as the *Financial Times* put it. Turning oil into chemicals may avoid the carbon impact of burning it as fuel, but pumping out ever more plastic creates a whole new set of harms, including to the climate. That didn't seem to bother Saudi Aramco. "We see the world changing," one executive explained, and the company was adapting "in a way that we keep our market share." For Saudi Arabia, "the big picture imperative is to avoid being forced to leave barrels in the ground," one analyst wrote. "Better to turn these barrels into petrochemicals" than "to leave reserves stranded." If the fuel market was weakening and plastic sales showed promise, Saudi Aramco would follow the money.

In 2020, three years after Sadara's completion, the oil giant committed even more definitively to plastic. Saudi Aramco bought a 70 percent stake in the Saudi Basic Industries Corp., or SABIC, a state-owned petrochemical conglomerate established by royal decree in 1976. SABIC had since grown into the world's fourth-biggest chemical company, with annual sales close to $35 billion—"one of Saudi Arabia's crown jewels," a regional news site wrote. Much of that growth came through gobbling up pieces of other companies, most notably GE Plastics, a General Electric division making plastics used in packaging, cars, consumer goods, and medical offices, for which SABIC paid $11.6 billion in 2007. Today, SABIC operates dozens of plants around the world, including a giant ethane cracker and polyethylene complex it built with ExxonMobil near Corpus Christi, Texas. Indeed, the company has more than a dozen manufacturing sites in the United States.

"A FANTASTIC WINDOW OF OPPORTUNITY"

Saudi Aramco's $69 billion acquisition of a controlling stake in SABIC was, in a sense, just one arm of the state buying from another, since both companies were almost entirely government-owned. Still, it was a watershed moment. SABIC's petrochemical output was more than three times Saudi Aramco's, so the merger was a clear recommitment to the oil giant's all-in bet on plastic.

Not long before, Saudi Aramco had announced plans to pour $100 billion—not counting SABIC's price tag—into expanding plastic and petrochemical production. Nasser, the CEO, said he saw "great potential in nonmetallic materials as a substitute for natural materials" in "high-growth industries" including packaging, automotive, and construction. The world's all-but-insatiable hunger for such items was why petrochemicals would account by midcentury for nearly half the growth in global oil demand—and therefore, "provide a reliable destination for Saudi Aramco's future oil production," he explained. It was simple, another company executive said: "As populations grow they need more plastic." That dynamic, in Nasser's view, provided "a fantastic window of opportunity." But "such windows by their very nature offer maximum benefit only to those who act quickly." That's exactly what he planned to do: By the early 2030s, Saudi Aramco aims to turn more than a third of its crude into chemicals—a near-tripling in fifteen years.

Rolling out that $100 billion spend would now be SABIC's work too. It depends, in large part, on a new set of technologies with the potential to transform the way plastic is made. "Crude oil–to-chemicals" promises—or threatens, depending on your perspective—to supercharge production. "A game changer," one industry publication pronounced. "Because of its sheer scale and the volume of products that can potentially be produced," crude-to-chemicals "could be one of the most important trends ever to impact" the industry. We've seen how byproducts of gas—particularly ethane and propane—are "cracked" into ingredients for plastic. When the process begins with oil rather than gas, it works differently. Traditionally, refineries separate crude into its component parts, or "fractions." Some—like gasoline, diesel, and kerosene—are fuels. Others, particularly the fraction known as naphtha, become ingredients, or "feedstocks," in the industry's lingo, for petrochemical plants that turn them—through a series of steps—into plastic and other products.

Each step is expensive. And refineries' output is heavily skewed toward fuel, traditionally processing just a tenth of oil into chemical feedstocks. Many have begun tweaking their operations to nudge that proportion higher. Given its new priorities, Saudi Aramco wants to rewrite the equation more definitively. Already, working with Chevron and a Houston-based company called Lummus Technology, it has developed a heat-based technology soon to come into use at a complex under construction in Ulsan, South Korea, that's expected to turn 25 percent of the oil it takes in into chemicals. "Definitely a breakthrough," one expert opined.

Saudi Aramco's not satisfied. "What if the yield could be increased even more?" its website asks. Inside a curving, glass-walled research building at company headquarters in Dharan, scientists are working on ways to push chemical output to 80 percent of every barrel. "This is truly pioneering research," the lab's development manager said in an article featured on the corporate site. Teams are searching for a catalyst—"the elixir," the company calls it, that, when mixed with crude, would trigger reactions yielding streams of useful feedstocks. Among other work, the article describes lab experiments focused on maximizing output of ethylene and propylene—the building blocks of polyethylene and polypropylene plastics. "Imagine you are making food and the recipe has many ingredients," one scientist says. You keep adjusting the proportions "to get the desired flavor and texture."

At another Aramco lab, teams work toward what the company calls the holy grail: crunching the crude-to-chemicals process down to one step, which would slash costs. It "seems like science fiction," but is getting closer to reality, one scientist said. The company's sharp minds and disciplined execution, said another, boost its odds of "hitting the jackpot."

It's hard to overstate the impact such research could have. Refineries are gigantic, so shifting them away from fuel production could "easily" push chemical output ten times higher than a typical petrochemical plant's, one expert said. "It will only take a few of these projects to significantly affect global supply and demand." To flood the world, in other words, with vastly more plastic. If Saudi Aramco reaches its 80 percent target, one crude-to-chemicals complex could produce more ethylene than the eight crackers built in the first wave of fracking-driven US plant construction do collectively, Standard & Poor's analysts estimated. Even less optimized projects already in the works are likely to "reshape the global petrochemical industry," they wrote.

Saudi Aramco is moving ahead with plans for a big crude-to-chemicals facility beside the Persian Gulf and possibly another on the Red Sea. Such complexes could double the money a typical refinery-plus-petrochemical-plant setup makes per barrel.

They also lock in demand for oil. Which is a big part of why Saudi Aramco, collaborating with Chinese companies, is now building, or buying into, major crude-to-chemicals complexes in that country. Among other ventures, it paid $3.4 billion in 2023 for a 10 percent stake in the corporate parent of Zhejiang Petroleum and Chemical, which operates a vast refinery whose output is about 40 percent chemicals—including more than four million tons of ethylene a year. As part of the deal, the parent company agreed to buy nearly five hundred thousand barrels of Aramco oil a day. Now Aramco is pursuing similar stakes in Jiangsu Shenghong Petrochemicals and Hengli Petrochemical, which run massive refinery and chemicals complexes, and Shandong Yulong Petrochemical, which is building one. If those deals are finalized, they'll likely include oil supply commitments too. Meanwhile, SABIC is building a $6.4 billion ethylene, polyethylene, and polypropylene plant in Fujian province.

PLASTIC INC.

Of course, Saudi Aramco isn't the only one tempted by the returns crude-to-chemicals promises. Across the industry, expected declines in demand for oil-derived fuels threaten to leave refineries running below capacity, McKinsey consultants explain. Operators, they advise, "must fundamentally rethink" their setups and move toward petrochemicals, either by adjusting existing processes or installing crude-to-chemicals technologies. Many are already doing so, said a contractor working with such facilities. "Any time we look at new refineries" or upgrade existing ones, "we almost always find that a major goal is petrochemical production," he said. Facilities are introducing it "at a scale that we really haven't seen before."

DRAGGING MY WHEELIE BAG THROUGH DUBAI'S GLITZY, GLIMMERING AIRport, I look up to see a huge ad promoting Saudi Arabia—its stunning landscapes and millennia of history—as a tourist destination. I'm taken aback by the idea of vacationing in a country I've associated more with the brutal oppression of women and exploitation of migrant workers than sightseeing potential. But given the reason for my trip to Dubai, the biggest city in Saudi Arabia's next-door neighbor the United Arab Emirates, the pitch makes sense. Attracting visitors beyond the millions making the annual hajj pilgrimage is part of Saudi Arabia's effort to develop new sources of income. The same forces behind its push to make more plastic are driving the development of luxury resorts on the Red Sea.

Plastic producers have frequently replied to my requests to interview their representatives with, at best, written statements, and more often refusals to comment, or just silence. (Neither Saudi Aramco nor SABIC responded to my messages.) So I was pleasantly surprised when organizers of the Gulf Petrochemicals and Chemicals Association's annual plastics conference said I was welcome to attend. SABIC was the main sponsor of the conference, whose other participants would include representatives of both global and regional petrochemical companies, so the gathering seemed like a good place to get a glimpse of Saudi, and Gulf, plastic plans.

Dubai feels astonishing from the moment I step into the enormous airport, with its mirrored columns, shiny marble floors, and walkways lined with tall palm trees. The diversity visible on the city's streets—and reflected in the restaurants I hungrily explore—is evidence of the economic draw it offers to workers from across South Asia, Africa, the Middle East, and Europe. The preserved, refurbished old quarter retains hints of Dubai's past as a small-scale hub for fishermen, traders, and pearl merchants.

But in much of the rest of the city, the UAE's wealth is on flashy, over-the-top display. In some neighborhoods, nearly every building is an architectural extravaganza—one neck-craning luxury tower after the next gleaming in the broiling desert sun, some garish, others elegantly avant-garde. As my Uber approaches the conference hotel, I can see it is another such marvel. Two, in fact—side-by-side oval-shaped buildings, connected fifty stories up by a walkway jutting out on one side into a cantilevered observatory with see-through floors. Visitors, I'd later learn, could zoom down a slide—outside the building, although enclosed in glass—from level fifty-three to fifty-two, or, even more harrowing, walk on an outdoor ledge, and climb the side of the tower, protected only by a harness.

Despite the buildings' lavishness, the plastics meeting is in a downstairs space that's less Dubai bling than basic corporate nice. The heavy Saudi presence is easy to see as I enter the refreshment area, where men in the country's traditional dress—long white robe, red-and-white-checked headdress secured with a dark band—chat over coffee and pastries with attendees in suits. At a SABIC information stall, several bearded young Saudis peer at a laptop, beside a tray piled with chocolates wrapped in gold and silver paper. They smile as I pocket a couple and walk into the conference hall.

On the screen up front, a video touts plastic's many benefits, notes its production has doubled since the century's start, and offers the idea of a "circular economy"—a world in which discarded material is endlessly recaptured and recycled—as the answer to any problems such growth might pose. "The time is now to harvest the power of innovation and build a new future for plastics," the narrator says. It's the first of many such exhortations that will soon pour forth in a blur of PowerPoint presentations and panel discussions.

Over a day and a half, representatives of Saudi corporate and government entities—including SABIC itself, a company-backed economic development initiative, and a Ministry of Energy official—take the stage, along with other petrochemical executives, representatives of big plastic users like Pepsi and Unilever, and trade groups including the US-based Plastics Industry Association. As the upbeat corporate-speak washes over me, I begin to imagine playing bingo with the most frequently repeated greenwash buzzwords—innovation, collaboration, opportunity, challenge, sustainability, circularity. Procter & Gamble, found in one waste audit to be the world's seventh-biggest corporate plastic polluter, is committed to "creating products that make responsible consumption irresistible to our consumers," one of its executives says—twice. "I'm optimistic because we have all of you," he adds. "The journey is probably still at the start," but it's time to "join hands together to achieve this." A packaging manager from Dow Chemical talks about its goals: "to protect the climate, stop the waste, and of course to close the loop."

Despite the sometimes anodyne language, there are moments when it seems impressive, and unexpected, to hear big corporate interests recognize the urgency of our planetary crises and the need to change. Packaging "is very, very important for us," a Unilever executive says. "But we know there is a problem." Then, unwrapping one of the SABIC chocolates, I remind myself who is paying for this event: the huge chemical conglomerate through which the Saudi state plans to monetize its oil reserves far into the future. Neither SABIC nor its partner, Saudi Aramco, currently pouring $100 billion into their petrochemical dreams, is interested in making less plastic. They are, though, offering a master class in industry's long-standing skill at co-opting the language of environmentalism.

Indeed, when I listen more closely, there is a lot more going on amid all that green talk. A "leadership dialogue" with three Gulf-based petrochemical executives hits many of the industry's favorite points. As the panelists settle into armchairs onstage, a moderator says there is a "difference between plastic pollution and plastic production. Plastic is good, pollution is bad." Turning to climate change, he shows a graph of rising emissions "the world cannot afford to ignore." As a component of wind turbines, solar panels, and electric vehicles, he argues, plastic will be key to shifting toward clean energy.

Fayez Al Sharef, CEO of Sadara, the massive Aramco-Dow petrochemical project, picks up the theme. "We're all citizens of this world, and we're affected," he says, his demeanor assured and thoughtful, black-framed glasses and close-trimmed beard beneath his checked headscarf. "We have to be part of the solution." But his industry is up against a troubling obstacle, he explains. People get "too emotional" about plastic. "We have this tendency to go between extremes. We want plastic, and then someone shows us a video about the waste, and the next day we say we don't like plastic." He takes a similar tone on climate. Shifting toward clean energy "has to happen. But not at the pace people are talking about, and not in the exclusion" some suggest, he says. "We need all forms of energy and we need to be patient." Hinting at qualms that buck this event's green hue—"I have to be careful"—he notes that fossil fuels provide plastic's raw materials. "I'm just wondering, with all these aggressive policies" on climate, could problems with availability of needed ingredients result? Some components of an oil barrel become fuel while some go into chemicals, "so if you disturb part of it without paying attention," you risk a domino effect of disruption. Crude oil–to–chemicals, he suggests, could help solve that problem.

Such messaging needs unpicking. The moderator's argument that plastic is a problem only if it ends up in the wrong place is a long-standing industry framing—the same message Keep America Beautiful has espoused—that cleverly accepts public concern while narrowing its target. As long as bags, bottles, and such are properly disposed of, the thinking goes, production can keep climbing. Talk of plastic's role in reducing emissions is common too. It's not wrong—the material does make vehicles lighter and less fuel-hungry, and it's needed in many clean technologies. But talking up those benefits, like reminders of plastic's role in medicine, is a way of focusing attention on its positive uses while ignoring the many wasteful ones. Arguments for an "all of the above" approach to energy—keep burning fossil fuels while also installing solar panels—and suggestions change will be slow are common industry talking points on climate, combining a reasonable-sounding tone with disregard for the urgency the danger demands.

As the day goes on, recycling is a constant theme. "It is crucial that we move from the traditional take-make-dispose economy to a circular one," says Jacob Duer, president of the Alliance to End Plastic Waste, an industry group whose members include major chemical and packaging makers. This is another familiar industry go-to. Recycling, and accompanying notions like circularity and "closing the loop," are plastic makers' answer to nearly every concern about their industry, despite the daunting technical, logistical, and economic obstacles, and environmental harms of their own. What's more, talk of recycling often focuses on "education" and "behavior change"—a handy way of shifting responsibility from industry to individuals.

The spin is even more shameless at a conference dinner, where young entrepreneurs competing for funding from the Mega Green Accelerator, a SABIC- and PepsiCo-backed start-up support program, present their plans: renting solar panels to small-scale farmers ("as of today, we've deployed ten units"), solar screens that offer both power and shade, an AI tool for calculating carbon footprints. Excellent projects, no doubt. But the idea that

"A FANTASTIC WINDOW OF OPPORTUNITY"

throwing a little money at them might begin to erase the environmental harm wrought by two vast global conglomerates is beyond absurd.

Sometimes the mask of concern slips, revealing flashes of anger. Few things irritate plastic makers more than the suggestion the world needs less plastic, so a briefing on negotiations for a global plastic pollution treaty, where they've been trying to bat that idea down, is one of the spiciest sessions. Salman Alajmi, an executive at the Kuwait-based petrochemical company Equate, who's been leading a coalition of plastic-producing countries at the UN talks, reports sentiment is "getting very emotional against plastic."

The treaty, Alajmi explained, was supposed to focus on "mishandling of waste—which is, again, consumer behavior." Demands for production cuts strayed far beyond those parameters, he complains. Some were even arguing that the calculations of a plastic item's impact should begin with oil and gas extraction, meaning the harm done by a throwaway fork, for example, would include not just its manufacture and disposal but the drilling and petrochemical processing that created the plastic. Alajmi warns that could pave the way for financial penalties that "will diminish for sure the producer economics."

If industry did not succeed in reshaping the deal, Alajmi warns, the consequences could be serious. "We have to be more proactive" and "get out of the comfort zone, out of the defensive mode." Pro-industry negotiators need legal experts, and research papers on the benefits of different types of recycling—"why they're safe and why we consider them solutions." It all sounds very concerning, says the moderator. "This is an invitation, and a call to action."

With time to kill between sessions, I follow a glass-enclosed footbridge over traffic-clogged roads to the Dubai Mall—the world's largest, with more than 1,200 stores. Wandering its huge, interconnected wings and walkways, I begin worrying I won't find my way back and decide to head for the indoor waterfall. Instead, I end up outside, by an artificial lake at the foot of the Burj Khalifa, the world's tallest building, whose jagged, asymmetric uppermost segments rise toward a sky-piercing needle. On a shaded terrace, I order a lemon-and-mint drink and am scrolling my phone when loud music begins to play, and dancing jets of water, in shifting patterns, spray up from the lake. At the world's largest mall, beneath the world's highest building, I have stumbled—in the middle of the desert—onto the world's tallest performing water fountain.

Trudging back through the mall, I stop to look down from its fourth floor at the multitude of stores below. It is dizzying. Dubai reminds me, I decide, of Las Vegas—minus the gambling, which is forbidden by Islam. Las Vegas on steroids. Or rather, on oil money. While the UAE's economy is far more diverse than Saudi Arabia's, oil is nonetheless the foundational source of its wealth. Like Saudi Arabia, the UAE, too, is pushing into plastic. To be in a place built—so very expensively—on fossil fuels, listening to endless green talk from an industry profiting off planetary destruction, is far more discombobulating than the astonishing shopping options. More head-spinning, even, than I imagine climbing the outside of the conference hotel would have been, had I been bold enough to try it.

ANOTHER WORRY THAT KEEPS COMING UP AT THE CONFERENCE IS CHINA. IT has long been the world's largest buyer of plastic, which its factories consume in huge volumes to make goods—appliances, toys, shoes, clothing—sold around the world. But in recent years, the country has grown wary of depending on others for a material so central to its economy. In addition to being the top plastic importer, it's the top producer too, and now it's ramping up to make even more of the material at home. Unlike Saudi Arabia's, this expansion is motivated not by a need to diversify but by a desire, in a time of rising nationalism, to become more self-sufficient.

As with so much of what China does, its petrochemical buildout has been mind-boggling in scale. Over five years, it has added as much production of ethylene and propylene—the two most important petrochemical building blocks—as exists in Europe, Japan, and Korea combined. In little more than a decade, China nearly doubled its output of high-density polyethylene and tripled its capacity to produce paraxylene, a precursor for polyester and PET, the plastic used for bottles. That growth's size and speed, reported the International Energy Agency (IEA), "dwarfs any historical precedent," roughly doubling the pace the Middle East and the United States previously hit.

On Dayushan Island, near Shanghai, Zhejiang Petroleum and Chemical's massive refinery—the crude-to-chemicals "megaproject" Saudi Aramco bought into—opened in 2019, then doubled its output three years later, to an astonishing sixteen million tons of chemicals annually. Hengli Petrochemical—which Aramco is also eyeing a piece of—began in the 1990s as a small polyester-weaving outfit, then grew into one of the world's largest producers of the synthetic yarn. Now it has an enormous refinery in the city of Dalian, churning out millions of tons of paraxylene. Hengli processes it into purified terephthalic acid, or PTA, converting that in turn into PET plastic and polyester fiber. Other chemicals pour from the complex too, sold off if Hengli doesn't want them. The company's expansion into making the materials it needs instead of buying them from someone else reflects the path China itself has traveled—starting decades ago with clothing factories, moving into polyester production, and then pushing deeper into the supply chain to synthesize the material's ingredients in petrochemical plants and refineries.

For those who were banking on selling plastic into China, that growth is a problem. Such companies "need to rethink their business model," an executive from NATPET, a Saudi polypropylene producer, warns from the Dubai stage. "We grew in our industry to become dependent" on Chinese customers, says Al Sharef, the Sadara CEO. Now, it's time to get used to "the new world."

But, Al Sharef notes, China's decision to make more of its own plastic has created opportunities too. The fundamental obstacle to its dream of self-sufficiency is a lack of adequate oil and gas reserves. So even in making its own plastic, it remains dependent on raw materials from abroad. And those with oil to sell are "looking for ways to work with China," Al Sharef says. "So it's not everything bad news." Indeed, Aramco is locking in its piece of that business already. China's petrochemical hunger, though, is bigger than one company. In fact, it's so vast that it's driving a realignment of global oil markets—keeping demand climbing even as sales of oil-derived fuels flatline. In late 2023, overall worldwide oil consumption bounced past pre-pandemic levels, climbing a million barrels a day higher than it was before Covid. Without Chinese petrochemicals, the IEA said, the total would have stayed well below 2019's levels.

It's not just oil. China is buying a vast and growing stream of fracked American feedstocks too. "Huge volumes of US ethane and propane have poured" in, the IEA said. "This symbiosis," it explains, "has enabled the petrochemical sectors in both countries to flourish in a way that would not otherwise have been possible." While China gets the largest share, American frackers are also supplying plants in India, Mexico, Brazil, and Canada.

So, at bottom, China's drive for self-sufficiency simply serves to underline the interconnectedness of the global petrochemical web, a network whose cross-border partnerships and complex supply chains themselves reflect the long-established patterns of the industry plastic making is built upon: fossil fuels. All the shifting of geopolitical winds, and the vagaries of new technologies, may change where, and how, plastic gets made. So far, though, one thing has stayed constant—the ever-climbing trajectory of its growth.

CHAPTER 6

Invisible Poisons

1982—Dr. Barney Clark receives the first permanent artificial heart, made mainly of polyurethane, at the University of Utah. Seventy-eight million metric tons of plastic produced.

Theo Colborn was always a bird-watcher, and had loved nature and science for as long as she could remember. But a career studying animals would have been unusual for a girl growing up in the 1930s, and Theo—it was short for Theodora—became a pharmacist instead. She raised her four children on a sheep farm in Colorado and then, pondering at fifty what her next phase might be, considered filling prescriptions again. Instead, she followed the passion she'd never forgotten, and began working toward a zoology degree. Some of her advisers weren't sure a middle-aged student was worth their time, but she got her PhD when she was fifty-eight, in 1985. And while her research began with wildlife, it would have profound implications for the health and well-being of humans too.

In 1987, scientific studies were piled high on every surface in Colborn's Washington, DC, office and stuffed into boxes that covered the floor. Her boss at the Conservation Foundation, a research group, shoved more papers under her door every day, and it was impossible to keep up.

She was part of a project on industrial pollution in the Great Lakes, a blight that had improved dramatically since the days when rotting algae covered beaches and Lake Erie was declared dead. Even so, the research Colborn was sifting through held some worrying findings. There were cormorants with twisted bills, missing eyes, and club feet. Seemingly healthy tern chicks inexplicably wasting away. Eggs that failed to hatch, and male gulls with female cells in their testes. Her boss had asked her to look for cancer in fish—it wouldn't have been surprising given the decades of contamination—but what she saw instead was more mystifying.

Trying to make sense of it, Colborn went back to her boxes, which now held more than two thousand scientific papers. Certain there must be a pattern in the puzzling symptoms, she tracked them on a spreadsheet, cross-referencing species like otters, coho salmon, and snapping turtles with problems including suppressed immunity, behavioral changes, and impaired reproduction. One thing was quickly clear. In most cases, adults were doing fine. It was the young who were suffering.

To Colborn, most of their symptoms seemed to reflect development gone awry. In wildlife and humans alike, it is the endocrine system—chemical messengers known as hormones, the glands that secrete them, and receptor cells around the body that respond to them—that governs development. Traveling through the bloodstream, estrogen, testosterone, insulin, adrenaline, thyroid hormone, and dozens of others guide not only sexual maturation and reproduction but also physical growth, brain development, metabolism, blood sugar regulation, stress responses, and other processes. It's an intricately balanced network, sensitive enough to respond to hormone level changes equivalent to one drop in twenty Olympic swimming pools. Perhaps, thought Colborn, chemical pollutants were interfering with this extraordinary system. Hanging over that emerging hypothesis was a terrifying question, not yet articulated: If contaminants were disrupting hormonal communication in birds and fish, could the thousands of chemicals in commercial use be wreaking similar havoc in humans?

Colborn didn't know much about the endocrine system, so she bought a medical textbook and flipped through it as she worked. With this new lens, Great Lakes creatures' problems began to make sense. The idea that artificial chemicals might be mimicking or blocking hormones was provocative. Scientists' traditional understanding of toxicity had focused mostly on cancer, with chemicals typically considered safe if they didn't cause tumors or obvious birth defects. The notion that another mechanism might be causing different—and harder to spot—harms would be a paradigm shift.

It wasn't without precedent. In the 1930s, Northwestern University researchers had dosed pregnant rats with estrogen. Tinkering with a mother's hormones, they found, could profoundly affect her offspring—the pups displayed serious abnormalities, including deformed vaginas and uteruses in females and stunted penises in males. It wasn't just reproductive organs that were affected. As a fetus grows, hormones help organize brain pathways that are soon set permanently (the hard-wiring process continues through childhood). They also influence development of the immune system, the body's overall growth, and maturation of organs including the lungs. For all those systems, normal development depends on the right signals getting to the right place at the right time, an "elaborate chemical ballet" conducted at "a dizzying pace," Colborn later wrote, with two coauthors. "If something disrupts the cues" at a critical point, "it can have serious lifelong consequences."

That had been made painfully clear with the fallout from a drug known as DES, which acted like natural estrogen and was prescribed to millions of pregnant women in the (sadly, false) belief it prevented miscarriages and premature birth. It was first synthesized in 1938, but only in the '70s—when a Boston hospital diagnosed a cluster of young women with an extremely rare vaginal cancer—did scientists realize it harmed those exposed in utero. On top of the unusual cancer, DES daughters suffered elevated rates of infertility and breast cancer. Their experience, Colborn saw, offered important lessons. For one thing, it demonstrated problems baked in before birth were not always visible, nor immediate. It showed a substance that did little harm to adults could have serious, life-changing effects on a developing fetus. And it made clear the human body could mistake a chemical produced in the lab for a hormone. She couldn't help wondering whether, if it hadn't been for that cluster of unusual cancers, anyone would have linked young women's health problems to a drug their mothers took decades earlier. It left her, by implication, with another worrying question. Could other chemicals be causing hormonal harms that were not yet understood?

One already known to be dangerous was the pesticide DDT. Long after Rachel Carson's landmark book *Silent Spring* spotlighted its harms, scientists discovered it did much of that damage by mimicking estrogen (while the chemical it degraded into acted like testosterone). Among its effects was a thinning of bald eagles' eggshells so severe, their young perished. Humans were affected too. Those exposed in the womb had higher-than-normal rates of breast cancer, hypertension, and obesity. A 2021 study in which researchers defrosted decades-old blood samples to test for DDT found the *granddaughters* of women with high levels in pregnancy started their periods earlier than average and were more than twice as likely to be overweight or obese—both warning signs for health problems later in life. Girls are born with all the eggs they'll produce, so changes may have been written into the granddaughters' futures while their mothers were in the womb. It was a frightening indication that endocrine disruption's harms can span three generations.

One strange thing was that while DES and DDT mimicked estrogen, neither much resembled it chemically. Yet they seemed to bind with estrogen receptors, cells around the body programmed to grab on to molecules of the hormone. Such receptors—there are hundreds of types, each matched to a particular hormone—work a little like locks, into which hormone molecules fit like a key. Once unlocked, the receptor carries out the hormone's instruction, switching a particular biological process on or off. It's all orchestrated by the hypothalamus, in the brain, and the pituitary, which together monitor blood levels and fire off hormonal messages telling glands such as the adrenals, pancreas, thyroid, ovaries, and testicles to crank their own hormonal output up or down.

What the DES and DDT tragedies demonstrated was that synthetic chemicals can trick the receptors into responding to them as if they were real hormones—even when their molecules look completely different from those of the substance they're aping. An estrogen receptor, Theo Colborn and her colleagues explained, "is a lock that can be opened with devices that bear as little resemblance to natural estrogen as a hammer does to a key. Even more puzzling, a wrench might work as well as a hammer." With so many chemicals already out in the world, the idea that a wide variety might be able to act on the body like estrogen did was deeply worrying. Colborn was especially concerned by a talk in which a Swedish expert pointed out toxicologists' ability to test for the presence of specific chemicals was falling ever farther behind industry's production of hundreds of new ones each year. Biologists reporting a reduction in the size of male fishes' testes in the Baltic Sea couldn't figure out which pollutant to blame, because they could identify only 6 percent of the suspects. This pair of problems—vast, ever-increasing numbers of chemicals, and companies' ability to shield as corporate secrets the most basic details about their identity and potential dangers—would dog efforts to understand their impacts for decades to come.

MANY HORMONE-DISRUPTING CHEMICALS, IT TURNED OUT, WERE INGREDI-ents in the plastics whose presence was already so inescapable. That worrying truth came into focus in 1987, when Tufts University researchers

Ana Soto and Carlos Sonnenschein were studying human breast cancer cells, experimenting with a strain that multiplies in the presence of estrogen. They were surprised when cells they hadn't juiced with the hormone suddenly began reproducing as if they'd gotten a dose. The lab's protocols were meticulous, so it was hard to imagine how their samples could have been contaminated, but Soto and Sonnenschein went over the possibilities again and again. It was months before they realized that storing their experiment's ingredients in a different brand of test tube prevented the unexplained estrogen response. Pressing their original tubes' manufacturer, they learned the company had recently changed the composition of its plastic. Something in the new formulation, it seemed, was mimicking estrogen's effects. It took months more work to identify the chemical in question—called nonylphenol.

Plastics are made up of not just the main building blocks—chemicals like ethylene and propylene—but also additives mixed in to give the end product desired properties. Nonylphenol is an antioxidant, used to prevent oxygen from weakening some plastics. Additives have many other functions: Plasticizers turn rigid raw material soft and flexible; pigments add color; stabilizers stop light or heat from damaging plastic; flame retardants prevent it from catching fire; fillers bulk it out. "Making things out of plastics is like playing a game with molecules," one industry group explains, and additives are key to that process. They can account for nearly half of a finished plastic's weight.

Because they're not tightly bound to a plastic's backbone molecules, additives break away easily—particularly when heated. Some are attracted to fats, so ooze into food or drinks from packaging. They can also enter the human body through skin, or when we breathe in microplastics floating in the air.

One particularly worrisome group are phthalates—a class of more than sixty individual chemicals that typically function as plasticizers, making material pliable. Often but not exclusively used in PVC plastic, or vinyl, phthalates are ubiquitous—present in packaging, toys, shower curtains, diapers and sanitary pads, the coating of wires and cables, and much more, including nonplastic products like nail polish, shampoo, lotion, and makeup. They include not just estrogen mimics but chemicals such as DEHP, which interferes with masculinizing hormones like testosterone.

Another harmful category is bisphenols, particularly bisphenol A, or BPA, which binds with estrogen receptors in the body and causes harm even in very low doses. It's typically used in a hard, durable plastic called polycarbonate and in epoxy resins that line food and beverage cans; it's also present in plastic water pipes, the casings for devices like phones and laptops, clothing such as sports bras and athletic shirts—even dental sealants. Despite long-standing awareness of its dangers, and its removal from a handful of product categories, BPA is still among the world's most widely used synthetic chemicals, with seven billion kilograms—and climbing—manufactured annually.

Theo Colborn's concerns mounted with every study she read, and in 1991 she called a meeting in Racine, Wisconsin, of about twenty researchers from fields including endocrinology, immunology, and toxicology. All were seeing different pieces of the problem she had identified, and together they issued a statement. They used the name "endocrine disruptors" for the more than twenty-five chemicals (some were actually groups including dozens of individual substances) capable of interfering with hormonal networks. These endocrine disruptors shared some distinct properties—their effects could reach across generations and might be apparent only after exposed offspring had grown. And while toxicologists had long believed "the dose makes the poison," the harm wrought by these sneakier substances seemed to depend more on the timing of exposure than its magnitude. Unless their use was dramatically reduced, the scientists warned, "large scale dysfunction at the population level is possible."

PLASTIC INC.

RESEARCHERS AND GOVERNMENT REGULATORS BEGAN, SLOWLY, TO TAKE notice. But the conference's warning didn't grab the public attention Colborn felt this danger deserved. So she joined with chemist John Peterson Myers and journalist Dianne Dumanoski to author a book. Published in 1996, *Our Stolen Future* set out the threat in stark, detailed terms. Failure to curb exposure to endocrine-disrupting chemicals, the trio wrote, would pose "the danger of widespread disruption in human embryonic development and the prospect of damage that will last a lifetime."

These substances, they explained, "play by different rules" than typical poisons or carcinogens. They "are thugs on the biological information highway that sabotage vital communication. They mug the messengers or impersonate them. They jam signals. They scramble messages. They sow disinformation." Much, although not all, of that damage, said *Our Stolen Future*, seemed to take place during the earliest stages of life, before and shortly after birth, while critical systems were developing in often irreversible ways. Imagine what would happen, the authors suggested, if communications were disrupted during construction of a building, "so the plumbers did not get the message to install the pipes in half the bathrooms before the carpenters closed the walls," thermostats were permanently set to the wrong temperature, or a high-rise ended up with one elevator instead of eight.

The disruptors' potential dangers, the authors noted, tracked with real-world health trends, from an apparent increase in neurocognitive problems like attention deficit hyperactivity disorder to a steady fall (then uncertain, now well documented) in men's sperm counts. They titled one of their chapters "Altered Destinies" to underline the idea that some of the harms were not diseases but subtler changes—difficulty focusing or reduced fertility—that could nonetheless diminish quality of life. "These are effects that manifest themselves in many shades of gray rather than in the black-and-white distinctions made between health and illness."

Most frightening, the authors believed, was the dearth of information about how many endocrine disruptors might be out there. "No one has systematically screened the tens of thousands of synthetic chemicals created since World War II for such effects," they wrote. While scientists' understanding was improving, the team noted research had also "underscored our astonishing ignorance about the man-made chemicals that we have spread liberally across the face of the Earth and incorporated into every part of our daily lives." But while the evidence was incomplete, Colborn and her colleagues concluded, it had "a cumulative power that is compelling and urgent."

Our Stolen Future scored the splash Colborn had been hoping for, grabbing headlines and TV time. Al Gore, then vice president, hailed it as a sequel to *Silent Spring*, the book often said to have launched modern environmentalism. Others pooh-poohed the findings. An opinion writer called the work "one part pseudo-science, two parts hype, three parts hysteria."

While some of the pushback came from independent scientists unpersuaded by the book's arguments, much of it was seeded by industries making and using the chemicals in question. The Chemical Manufacturers Association (since renamed the American Chemistry Council) sent reporters a twenty-three-page handout rebutting the book's claims. Even before *Our Stolen Future* was published, the chemical manufacturers, the American Plastics Council, the Society of the Plastics Industry, and pesticide makers—whose products were also implicated—began discussing ways to counterattack. "Quite honestly, my first reaction is that this might be of genuine concern to women and the general public," a pesticide publicist wrote, but she predicted plastic was likely to bear the brunt of such a response.

For all the skepticism—honest and otherwise—at the time, decades later Colborn looks prescient. Not every fear expressed in *Our Stolen Future* was borne out, and some of its language feels overwrought. But the fundamental dangers it described have not only been proved real but have grown even more pressing as endocrine-disrupting chemicals have spread ever more widely.

But Dumanoski, one of the book's coauthors, put her finger on a nagging problem. A journalist who'd translated her colleagues' scientific expertise into readable prose, she had seen in her days at *The Boston Globe* how terms like *ozone hole* and *acid rain* had made abstract threats real to the public. *Endocrine disruptors*—or even the more comprehensible *hormone disruptors*—didn't pack the same punch. "It just doesn't communicate," Dumanoski said. "And nobody has come up with a really zingy, wonderful alternative term." That, combined with the inherent complexity of the chemicals' operation, may have been why, despite the worry the book generated, the danger didn't stick in people's awareness the way the trio had hoped.

Theo Colborn, who died in 2014, used to sign her emails "Onward!" Thanks in part to the push she provided, that's where the science on endocrine disruptors has moved in the decades since *Our Stolen Future* came out. In 2006, ten years after publication, Colborn said accumulating research had made clear "we are in far deeper trouble than we ever anticipated while writing the book." By then, more endocrine disruptors had been identified, more hormonal systems had been found vulnerable, and more health conditions were linked to the chemicals than the authors had first imagined. And the number of chemicals being manufactured continued to grow. "Endocrine disruptors now are in your home—everywhere—everything you touch. They have become integrated into our lifestyle and our economy," Colborn said. "It would be great if we only had one or two chemicals to deal with. We have hundreds, now."

INVISIBLE POISONS

GIVEN THEIR UBIQUITY IN THE MATERIALS AND PRODUCTS WE USE EVERY day, it's unsurprising that endocrine disruptors have been found in the bodies of the vast majority of people tested. Today, the range of health problems linked to these chemicals is astonishing. And while BPA and phthalates remain two of the most troubling, there are many others—an alphabet soup of substances used to make plastics useful and appealing. They include PFAS "forever chemicals," whose contamination of soil and drinking water has exploded into a major public health concern in recent years. In addition to their presence in firefighting foams, cleaning products, tampons, and cosmetics, PFAS—a family of some ten thousand perfluoroalkyl and polyfluoroalkyl substances—are also used in plastics, including food and drink containers.

As has long been understood, endocrine disruptors take a direct hit on both the male and female reproductive systems. Elevated BPA levels are linked to premature ovarian failure and other fertility problems in women, as well as early menopause; phthalate exposure is correlated with endometriosis, a chronic and often debilitatingly painful condition. One study found pregnant women with the highest phthalate levels were 60 percent more likely to miscarry than those with the lowest. Another linked exposure to a near-tripling of the likelihood of premature birth and estimated phthalates were causing more than fifty-five thousand early deliveries a year in the United States alone. Endocrine disruptors pass through the placenta—they may even change its shape and size—and later taint a mother's breast milk. Boys exposed to elevated levels are more likely to have undescended testicles (a risk factor for testicular cancer) and a reduced distance between their anus and genitals, an indicator of lower fertility. Some phthalates are linked to decreased sperm counts and lower testosterone levels in adult men.

More surprisingly, the chemicals have also been tied to cardiovascular disease. One major study found people with the highest levels of BPA in their urine had a 73 percent greater chance of suffering a heart attack and a 60 percent higher chance of a stroke than those with the lowest levels. Another found it may contribute to kidney disease too. Chemicals in plastic also interfere with the body's metabolism, contributing to obesity, diabetes, and the prediabetic condition known as insulin resistance. Rodents exposed to endocrine disruptors in utero or as newborns were heavier and more likely to display insulin resistance later. And while Theo Colborn pushed scientists to consider non-cancer risks, it turns out hormone disruptors are linked to cancers too—particularly those of the breast, uterus, ovaries, and prostate. A particularly troubling pair of studies found the phthalate DEHP, a common ingredient in IV bags and medical tubing, stymies the effectiveness of some breast cancer drugs, increasing the chance of the disease returning and raising patients' likelihood of dying. The findings also suggested DEHP could switch on a gene that promotes the cancer's spread. "This may be the tip of the iceberg" on endocrine disruptors' interference with medical treatments, said Theo Colborn's old coauthor Pete Myers.

Perhaps most shocking are endocrine disruptors' effects on the brain. The research is still developing, but scientists have drawn connections tying ADHD and autism spectrum disorder—as well as lower IQ, behavioral problems, and reduced motor and communication skills—to prenatal, childhood, and adolescent exposure to the chemicals. One psychiatry magazine called plastics' impact "a neuropsychiatric problem hidden in plain sight." Endocrine disruptors, it noted, have been linked to anxiety, depression, and even potentially Alzheimer's and Parkinson's diseases.

It's hard not to notice that many of the conditions linked to these chemicals—or diagnoses of them, at least—are on the rise, particularly in young people who have grown up in a world saturated with plastic. So many children in my own circles are on the autistic spectrum or have ADHD, and the data supports my anecdotal impression that the numbers diagnosed have been increasing for years. Breast cancer and fertility problems, too, are climbing. There are many causes, of course, for the spread of such problems, from changing diets and lifestyles to improved detection and diagnosis. And there are other sources of endocrine disruptors, like cosmetics, pesticides, and industrial pollution. But plastics bring these chemicals into our homes and the most intimate areas of our lives—through packaging of the food and drinks we consume, medical equipment that touches inside our bodies, the toys our children play with, our shoes and clothes, and so much more. The idea that they are contributing to such frightening and mysterious trends feels destabilizing, shaking something foundational.

Tracing endocrine disruptors' harms is inherently difficult, given the complexity of their operation, the long lags between exposure and effect, and the near impossibility of finding a control group of unexposed individuals. Studies of their effects necessarily demonstrate correlation, not causation. Adding to the challenge of tracing such connections, and the potential peril, is the reality that, as Theo Colborn realized, these chemicals do their work in extremely low doses—like those we encounter every day. BPA, for example, "doesn't seem to have a threshold below which it doesn't cause any effects," one expert said. "I don't think there's a safe level." That makes us all guinea pigs in a giant global experiment most of us are barely aware of.

What seems clear, even with that experiment still underway, is that the impact is significant. Seeking to quantify it, one analysis attributed more than 350,000 heart disease deaths globally each year—including more than 13 percent among people aged fifty-five to sixty-five—to the phthalate DEHP. Another study estimated BPA was responsible for more than 7 percent of strokes and cardiovascular problems in the United States—meaning it causes 1.5 million cases of cardiovascular disease and nearly sixty-one thousand strokes every year. The same study put the cost of death, disease, and disability caused by three endocrine disruptors—BPA, DEHP, and a flame retardant called PBDE, used in mattresses, textiles, furniture's foam padding, and electronic devices—at $920 billion in the US alone. It's an enormous sum—greater than the GDP of all but the twenty largest national economies. And while that analysis focused on Americans, the mechanisms of harm are the same anywhere. The findings put "a bright, bold line underneath the fact that plastics are a human health issue," one expert said. "We're talking about effects that run the entire life span."

WHILE SCIENTISTS' UNDERSTANDING OF ENDOCRINE DISRUPTORS' EFFECTS has been advancing, the other trend Theo Colborn observed—the relentless growth in the number of chemicals being produced—has continued at breakneck pace too. One effort to inventory those used in plastic production found more than ten thousand, and said many had hardly been studied. Even so, it identified nearly a quarter as potentially harmful to human health. And with thousands of new chemicals coming onto markets every year, it's all but impossible for researchers to keep up.

Regulators are falling even further behind—and the United States is particularly lax. Its failure is foundational, stemming from the all-but-toothless 1976 Toxic Substances Control Act, or TSCA, which underpins chemical regulation. The law's ineffectiveness was no accident. "It was written by industry," Steven Jellinek, the first director of the Environmental Protection Agency division tasked with enforcing it, later recalled. After taking the job, Jellinek told an oral history interviewer, he learned at least five others had turned it down, convinced making the law work would be nearly impossible. Insiders called it the Heckert–Eckhardt bill: The names belonged to the Texas congressman who'd sponsored the legislation and a DuPont vice president who negotiated on behalf of the Chemical Manufacturers Association. The TSCA, Jellinek said, was "the antithesis" of powerful laws like the Clean Air and Clean Water acts, which were enacted a few years earlier packed with provisions designed to build in accountability and effectiveness. When it came to chemicals, there were "just all of these hurdles that had been put into the law to keep it from doing much."

Perhaps the most glaring was that the law did not require chemicals to be proved safe before being used in everyday products, as was the case for pesticides and medications. Instead, its "innocent until proven guilty" approach put the burden on the environment agency to demonstrate a substance was hazardous before it could ban or limit its use. If it tried to do so, it had to choose the "least burdensome" restrictions. For their part, companies simply needed to notify the agency of plans to introduce a new chemical ninety days before they planned to start producing it. The data they had to provide was so minimal regulators struggled to assess potential hazards, or even create tests to detect the new substance. And even much of that sketchy information could be concealed from the public by claiming secrecy was necessary for business reasons, an opacity that's made it hard for outside scientists to study chemicals' risks. On top of it all, the sixty thousand chemicals already in commercial use when the law passed were all presumed safe, a grandfathering that enabled the continued proliferation of substances later demonstrated to carry real dangers.

That industry-friendly setup explains why, in the four decades after the law's passage, the Environmental Protection Agency banned or widely restricted only five chemicals. The marquee demonstration of the toxics law's brokenness was that it took the United States until 2024 to ban asbestos—known for decades to cause cancer and long prohibited in dozens of other countries. The EPA began work on outlawing it at the end of the 1970s, with a ten-year, multimillion-dollar review process that concluded in 1989 with a decision to end nearly all commercial use of the material. Asbestos companies sued, and in 1991 a court struck down the ban. The judge criticized industry's "protest everything" approach—but nonetheless ruled regulators hadn't shown a ban was the least burdensome way to reduce risk to the public.

That decision hobbled chemical regulation for decades—gun-shy officials did not propose banning a single additional chemical for nearly thirty years. Fear of lawsuits had constrained regulators from the start, Jellinek said, recalling EPA attorneys' constant worry that courtroom losses would create precedents tying the agency's hands further. While frustrating, their caution was understandable, in his view. Chemical makers "had some of the best lawyers in Washington," and the agency "was being challenged on everything." Its weakness was baked in, but while many laws "get changed and made better" as their flaws become apparent, he observed, that "just hasn't happened" with the TSCA.

It was 2016 before the law was finally overhauled. Once again, industry helped shape it. As Congress worked on the bill, then-Senator Barbara Boxer said the coding of a version shared with her showed it had originated at the American Chemistry Council. "Maybe I am old fashioned, but I do not believe that a regulated industry should be so intimately involved in writing a bill that regulates them," she said. The council denied Boxer's claim, but while the version that ultimately passed improved on its decades-old predecessor, the industry got much of what it wanted.

One big win was a provision stopping states from enacting their own chemical restrictions, which some had done in the absence of federal action. On the other hand, the update removed the "least burdensome" language that had made any action so difficult. And it ordered regulators to begin testing the riskiest chemicals, at least twenty at a time (an earlier, unsuccessful legislative effort had sought to get thousands assessed). At that pace, it would take decades to evaluate all the chemicals already in use.

More fundamentally, the new law maintained its predecessor's "innocent until proven guilty" approach. And it continued to address chemicals one by one, rather than in groups of related substances. That flaw—one critic said it left regulators "emptying the sea with a teaspoon"—is bigger than just US law, stretching around the world and weakening public pressure campaigns as well as governments' rulemaking. It drives a phenomenon known as "regrettable substitution"—when manufacturers replace a chemical that's been outlawed, or gained notoriety through consumer activism or critical media coverage, with an untested, often equally harmful alternative.

That's what's happened with bisphenol A, or BPA. A groundswell of concern about its effects through the 2000s prompted the Food and Drug Administration to prohibit its use in baby bottles and sippy cups in 2012. Many states had already done so, and some companies voluntarily removed it from products beyond the baby aisle, then labeled reusable water bottles and such as "BPA-free." But often, they simply replaced the substance with other members of the bisphenol family—like bisphenol S and a chemical cousin known as BADGE—that are just different enough to be considered distinct. They're less well studied but appear to carry their own harms. Similarly, a study on the phthalate DEHP's links to premature birth noted chemicals used to replace it pushed prematurity rates even higher—some of them packing twice DEHP's punch. We all pay the price for what one scientist called "a never-ending shell game," even as companies boast of cleaning up.

What's more, regulators are perpetually outgunned by a chemical industry determined to keep rules as lax as possible. The updated toxics law, passed just before Donald Trump was first elected, in 2016, "hasn't been given half of a chance to succeed," then-President Joe Biden's chemicals chief later said. The first Trump administration, she noted, didn't request any additional funds to go along with the extensive new responsibilities handed to environment officials. "The conveyor belt is sort of stopping" on assessments of substances' dangers, one expert told *ProPublica*. It was no surprise. "The whole regulatory process is designed to be slow and to be slowed down by those opposed to regulation," another noted.

Leading that charge is the American Chemistry Council, which, in its earlier incarnation as the Chemical Manufacturers Association, helped shape the original toxics law. With more than 190 corporate members—including marquee names such as ExxonMobil, Shell, Chevron Phillips, Dow, and DuPont—the council's annual budget recently hit $180 million. It puts those vast resources to use relentlessly opposing any restrictions that might dent companies' profits. In the first half of the 2020s, as the Biden administration pushed to scrutinize more chemicals, the American Chemistry Council was consistently among the country's dozen biggest lobbying groups, often breaking the top ten (while its member companies put additional millions into individual influence efforts). Its political arm pours money into campaigns too, typically giving far more to Republicans than Democrats. (The group did not reply to my emails requesting an interview with one of its representatives. But it says it prioritizes the safety of the public, the industry's workforce, and the environment, and supports a "clear, consistent and transparent approach to determining and applying the best available science.")

The American Chemistry Council throws its weight around not just in Washington but across the country. It reported in one disclosure that it had "helped defeat, amend or postpone" more than three hundred chemical and plastics bills in forty-four states. A Connecticut proposal to require companies to disclose, and eventually remove, hazardous chemicals in products for children drew ferocious opposition. "It's an onslaught" by industry lobbyists, said the bill's sponsor. "You go into a committee and they're out here grabbing legislators one after another."

The council wields its influence even more directly when those who've spent years in its employ get their hands on government's rule-writing pens—the fox, to mix metaphors, taking charge of the henhouse. In the first Trump administration, for example, with fifteen of the top twenty environment jobs held by people from the fossil fuel, chemical, and agriculture industries, Nancy Beck jumped from the American Chemistry Council to become a top official in the environment agency's chemical safety office. Her rewrite there of standards for evaluating chemicals' hazards drew word-for-word on industry's proposals, a *New York Times* investigation found, and the lobbying group said that without the changes, companies would have lost "millions of dollars and years of research invested in a chemical." Beck, inevitably, was back at the regulatory agency for Trump's second term, as an even more aggressive assault on chemical regulation took shape. Her new Environmental Protection Agency colleagues included a lobbyist who spent more than thirty years at DuPont, another who shilled for PFAS makers, and a lawyer who'd challenged a ban on asbestos. As layoffs decimated the agency's staff, the new team was said to be working on a plan to change the framework underpinning limits on use of the dangerous forever chemicals in consumer goods, a move that could threaten not just federal but state PFAS rules too. Trump liked to repeat his health secretary Robert F. Kennedy Jr.'s promise to "Make America Healthy Again." But his administration's actions promise to do just the opposite.

THINGS ARE DIFFERENT ACROSS THE ATLANTIC, WHERE THE EUROPEAN Union takes a far more muscular approach to protecting its residents from toxic dangers. In 2007 it adopted a framework called REACH (it stands for Registration, Evaluation, Authorization, and Restriction of Chemicals), which has, in the years since, set a new global standard for controlling chemicals. While far from perfect, the EU stands out in a world awash in hazardous substances simply for taking "recognizable steps in the right direction," as a global endocrinology association put it.

"At least it's doing something," observed the British writer George Monbiot. Most fundamentally, Europe inverted the American "innocent until proven guilty" model. In theory if not always in practice, the EU required companies to provide detailed information on a chemical's potential dangers and, if there is evidence it might cause harm, prove it can be used safely.

In REACH's first decade and a half, the EU banned or restricted the use of about two thousand chemicals, including more than a dozen phthalates. Just one of its many orders—barring four specific phthalates from products including flooring, mattresses, and children's swimming gear—would annually save two thousand boys from impaired fertility later in life, Europe's chemicals agency estimated. It outlawed the use of BPA and three other bisphenols in any packaging or equipment that comes into contact with food or drinks, having already prohibited its use in bottles and food packaging for babies and small children.

REACH has its problems. It's slow and cumbersome, and regulators depend on companies to provide accurate data about their products. Its commitment to the precautionary principle—that chemicals should be proved safe before being used in consumer products—has often been "more myth than reality," one environmental group said. And while REACH allows large groups of chemicals to be regulated together, the truth is it hasn't often happened. So while Europeans suffer less toxic exposure than most of the rest of the world, they're still swimming in a sea of chemicals. A continent-wide screening effort conducted by more than one hundred government agencies, labs, and universities, analyzing blood and urine samples from people in twenty-eight European countries, found they were "exposed to alarmingly high levels" of many chemicals. While levels of specific, regulated substances had declined over time, both plasticizing additives and PFAS were found in all the children and adolescents tested.

That's why, in 2020, EU leaders announced plans to strengthen and update REACH, laying out a vision of a zero-pollution Europe where consumer products would be finally free of the most harmful chemicals. The overhaul was part of a wider push one environmental group hailed as a "great detox"—the world's largest ever chemical crackdown, which would take thousands of the most harmful substances out of use, including all bisphenols and all forms of PVC vinyl. Industry flooded officials with objections. With the populist right rising across the continent as the decade wore on, European politicians grew wary of angering big companies with tough regulations. Ambitions began to ebb, and a target date for finalizing the plans slipped further into the future. Years into the effort, one lobbying watchdog told me, little had been delivered, and there'd been "a lot of political backtracking." If still a step forward, it was shaping up to be much smaller than originally envisioned.

On the other side of the English Channel, ambitions have shrunk even more dramatically. When Britain's exit from the EU became official at the start of 2021, it also ended participation in REACH, which, for all its shortcomings, remains the world's strongest set of toxic chemical regulations. Little noticed in the political maelstrom of the post-referendum years, the rupture was of a piece with the "hard Brexit" Boris Johnson pushed through—a definitive break that loosed Britain from the continental regulatory state so loathed by the most ideologically committed Brexiteers. Like Switzerland, a fellow non-EU member, Britain could have decided to adopt new REACH rules by default, retaining the right to diverge whenever it saw fit. Instead, it chose to build its own system from scratch.

REACH was written into British law, and European regulations in effect at the moment of exit became British regulations. But in the years since, as the EU has moved forward with action on newly identified threats, UK REACH—one expert called it a "cut-price system"—has fallen further and further behind. Its overwhelmed and overworked civil servants simply can't manage the enormous job of evaluating the hazards of thousands of chemicals. "The EU REACH system achieves incredible economies of scale," Chloe Alexander, of the chemical safety group CHEM Trust, told me. The EU employs hundreds of staff with the necessary, hard-to-find skills, and draws on the expertise of government agencies across the continent. "We simply can't replicate that capacity," Alexander said.

It's all added up to a sharp—and ever-growing—cross-channel divide. In UK REACH's first four years, as the EU imposed restrictions on fifteen substances and began work on limiting two dozen more, Britain finalized no new restrictions and initiated proceedings on just three. Where it does regulate, the protections are typically weaker than Europe's. Over the same time span, as Europe added thirty-three chemicals to its "substances of very high concern" list—often a precursor to a ban—Britain added none to its list. The EU also updated its approach to classifying chemical harms, creating a new category for endocrine disruptors, an important step that will make it easier to outlaw them. Britain declined to follow suit, so its rules on hormone-disrupting chemicals stand still while Europe prepares, among other actions, to prohibit them from toys. Labour, which won power three and a half years after UK REACH's creation, has professed a desire to return to closer alignment with EU chemical rules. It's unclear whether, or when, such a change might come. Without it, Alexander told me, "harmful products" that don't meet EU standards will be "dumped on the UK market."

Brexit's true meaning is the subject of endless debate, but it doesn't seem many who supported it were voting for a looser stance on toxic chemicals—not knowingly, anyway. In a 2020 poll of younger Leave voters, more than 70 percent expressed support for keeping or increasing EU regulations on chemicals. Then–environment secretary Michael Gove, one of Brexit's chief architects, promised in 2018 that "not only will there be no abandonment of the environmental principles that we've adopted in our time in the EU, but indeed we aim to strengthen" protections. It was one of many Brexit promises that would never be fulfilled. Indeed, Monbiot, the journalist, saw UK REACH's shortcomings as a "failure-by-design" that reflected the anti-regulatory zeal of Leave's hardest-core ideologues and wealthy funders, if not the majority of its voters.

Among the substances the EU has outlawed as Britain lags are plastic granules spread onto artificial-grass sports fields to make synthetic turf feel more natural underfoot. "Rubber crumb" used for the purpose often comes from shredded tires—which themselves are made from a complex cocktail of heavy metals and other hazardous chemicals never intended to be dived onto, slid along, or stomped into the air to be inhaled, as athletes often do when playing rugby, football, baseball, or other sports. A single field can require forty thousand tires' worth of rubber crumb, and needs to be freshened up periodically with more. In addition to polluting nearby soil and water, the toxic particles, some scientists worry, are causing cancers in those playing on the fields (the turf's blades and backing can contain PFAS, another potential source of illness). There's not enough research to know for sure, and synthetic turf companies say their product is safe. Europe's athletes—kids in weekend leagues, professionals pulling hefty paychecks, and everyone in between—will enjoy a measure of protection as that question is answered. Neither Britain's nor America's players will be so lucky.

WHILE EVIDENCE ON THE DANGERS OF ENDOCRINE DISRUPTORS AND other chemicals in plastic has been accumulating for decades—thanks in part to Theo Colborn's pioneering work—another equally worrying area of research has emerged much more recently. It's focused on microplastics, the tiny, physical fragments created as larger plastics break apart. As scientists have begun searching, they've found these particles almost everywhere they've looked. A single microplastic can be as big as five millimeters or as small as one micrometer—one one-hundredth the width of a hair, and able to penetrate individual cells.

The sheer scale of their presence and reach is almost impossible to grasp. One study estimated microplastics equivalent in weight to three hundred million bottles fall every year, via rain and wind, on an area that's about 6 percent of the United States' landmass. Extrapolated nationally, that suggests five billion bottles' worth of plastic particles are drizzling down onto the country every year. Most estimates don't include microplastics' tinier siblings, nanoplastics—smaller than a micrometer—which are far more numerous but harder to detect. One analysis that did account for them reported about two hundred billion nanoplastics were landing on every square meter of a remote corner of the Austrian Alps each week—more than a billion an hour. The particles are so omnipresent, journalist Matt Simon wrote in his 2022 book, *A Poison Like No Other: How Microplastics Corrupted Our Planet and Our Bodies*, that "plastic is now a fundamental component of the air." It's true of water too: In northwestern England, researchers found as many as half a million microplastics in every square meter of river sediment, and a team in India estimated the Ganges River was dumping three billion daily into the Bay of Bengal.

Micro- and nanoplastics are not only all around us—they are also inside us. By the early 2020s, as a trickle of research turned to a flood, scientists were reporting their presence in more and more parts of the body. They've now been detected in human hearts, lungs, bone marrow, livers, kidneys, intestines, penises, and testicles; and in blood, urine, semen, and breast milk (75 percent of samples in one study). Microplastics have even made their way into the placentas cocooning babies before birth. Stool analyses suggest most adults consume millions of microplastic particles a year, and the numbers are even higher for children.

For most of his career, Francesco Prattichizzo, a diabetes and heart disease researcher in Milan, wasn't particularly focused on plastic. But he'd long known that some proportion of the ailments he studied—perhaps a fifth of heart attacks, he tells me—couldn't be explained by known risk factors like poor diet, lack of exercise, genetic predisposition, or high cholesterol. That was driving many researchers to examine the role of environmental hazards, as they sought to better understand the origins of disease. So Prattichizzo was intrigued when a colleague, Raffaele Marfella, asked him to help with a study that might begin to answer the most worrying question microplastics' increasingly well-documented presence in the body raised: What might those tiny particles be doing to our health?

Doctors treating patients with severe blockages in their carotid arteries—which run up either side of the throat, carrying blood to the brain—sometimes remove fatty plaque from the vessels' walls, to reduce the risk of a piece breaking free and causing a stroke or heart attack. Working with doctors at several Italian hospitals, Prattichizzo tells me, the research team "had this idea: 'Why don't we try to look for microplastics'" in excised plaque. That's exactly what they did.

What they saw was startling enough to be published, in 2024, in *The New England Journal of Medicine*, among the most prestigious of scientific journals. Analyzing plaque samples from 258 patients, they found microplastics or nanoplastics in nearly 60 percent. The scientists took photographs with a powerful electron microscope, and the jagged, irregular shapes and opaque borders of the particles they captured made clear they were nothing natural. "We were all quite shocked" by the images, Prattichizzo says.

Even more alarming was what they learned next. As doctors monitored the patients for three years following their procedures, those whose arterial plaque had contained plastic were more than four times as likely as those without to suffer a heart attack or stroke, or to die. The findings—shocking in both their scale and the visceral punch they packed—made headlines well beyond medical journals.

While the results showed a clear correlation, it was too soon to know whether the tiny particles had caused the heart attacks, strokes, and deaths—and if so, how. Their presence might have weakened plaque, making pieces more likely to dislodge and enter the bloodstream, to create a dangerous blockage elsewhere. The body might have responded to the fragments with inflammation that set off a dangerous biological domino effect. Or any of the thousands of chemicals in the plastics—either endocrine disruptors or other, unidentified substances—might have been responsible. Causation could have worked the other way. Perhaps there was something about the worst plaque deposits that drew in more microplastics, and the particles were a symptom, not a cause, of a more dangerous problem.

Even with those caveats, the findings sounded a clear warning. More research is urgently needed, Prattichizzo says, to figure out exactly what harm microplastics might be wreaking in the body, how they're doing it, and whether there are specific steps patients could take—not microwaving food in plastic, for example, or avoiding drinks bottled in it—to reduce their exposure.

Such research is already well underway. And while the Italian team's findings were among the first to show a direct correlation with health problems, they wouldn't be the last. Just a few months after the plaque study appeared, a University of New Mexico group led by Matthew Campen, an environmental health expert, published similarly frightening findings.

They'd obtained brains from the autopsies of people who'd died in 2016 and 2024. Each brain contained, on average, about seven grams of plastic—roughly as much as five bottle caps or a disposable spoon. That on its own feels horrifying. "I never would have imagined it was this high," Campen said when the results were published. "I certainly don't feel comfortable with this much plastic in my brain." Even worse, his team found those who'd had dementia when they died had much more plastic in their brains than those without the diagnosis. Like the Italians, Campen's team could show only a correlation between plastic's presence and the incidence of disease, not a causal link. Still, it was a troubling portent.

One of microplastics' most worrying aspects is that each is a vehicle in which a bevy of different chemicals enters our bodies. In addition to carrying the chemicals used to make the plastic, they also pick up other pollutants—industrial toxins, pesticides, and more—as they travel through water or air. "Microplastic isn't a monolith but a many headed petrochemical hydra, a plethora of poisons," the journalist Simon wrote. Bacteria and viruses hop rides on them too.

Microplastics' ubiquity means we're constantly surrounded by them. Much of that exposure comes at home, where "almost everything around us is spawning microplastics," Simon wrote—synthetic fibers in carpets, furniture, and curtains; our shoes and clothing, from yoga pants to stretchy jeans, and even the sealants on hardwood floors. Every footfall or plop onto the sofa sends more into the air, so many that tens of thousands fall onto the typical living room floor daily.

As well as breathing the particles in, we also consume them when we eat and drink. Plastic bottles and containers shed microplastics into whatever they hold. But they're often present in food and drink even before it's packaged, picked up during processing—as milk is piped through plastic tubing at dairies, for example—or from water, air, or soil. One study found we could each be consuming about six thousand plastic fragments annually just in salt. Microplastics have been found in beer, honey, sugar, spices, rice, flour, and seafood. They penetrate many fruits and vegetables through the root systems of the plants that bear them. Heating plastic prompts it to release particles, so boiling water in a plastic kettle or brewing coffee in a plastic machine loads drinks with them. Disposable paper cups typically have an invisible polyethylene lining, so they shed microplastics too. Someone who drinks only bottled water might swallow more than 3.5 million microplastic particles a year, but even exclusively consuming tap water gets you nearly half a million annually.

Microplastics' origins are as wide-ranging as the ways they reach us. The trash that ends up in oceans and rivers, or on land, is perpetually fragmenting into particles. Another big source is car tires. They're made from both natural and synthetic rubber—the latter being a complex plastic. Driving's heat and friction causes tiny particles to fly off—the average tire loses about four kilograms over its lifetime. One estimate suggests the world's tires release enough microplastics annually to fill thirty-one giant container ships, each four hundred meters long. Blown by wind and carried by stormwater, those particles can travel long distances. Another unexpected source is paint—a great deal of which, based on acrylic latex, is essentially "pigment suspended in liquid plastic," as one writer observed. Tiny fragments enter the environment as paint wears.

Upward of one hundred thousand microplastic particles—if not many times that—drain away from every load of laundry. Some microplastics are produced on purpose—added to eyeliners, mascara, and lipstick, for example, to make them smooth and spreadable. Britain and the United States only prohibit the intentional addition of microplastics to products meant to be rinsed off immediately, like face wash, so manufacturers still put them in toothpaste, shampoo, sunscreen, nail polish, even dishwashing liquid. If they don't escape to the seas, they often accumulate in sludge that wastewater-treatment plants sell as fertilizer.

Microplastics' environmental damage is as worrying as their potential health harms. Through photosynthesis, algae remove carbon dioxide from the atmosphere and turn it into plant matter that sinks to the ocean floor, helping put the brakes on climate change by soaking up some of the emissions humans create. But scientists worry the microplastics now clogging oceanic vegetation could interfere with that process.

There's even evidence that the particles may be giving off climate-warming emissions themselves as they degrade. That research is at an early stage, but it adds to fears that microplastics could be significantly accelerating our climate crisis.

When I read about these particles' ubiquity and their dangers, I feel my anxiety rising—a mild panic that keeps me clicking through what feels like an endless chain of links, one headline more frightening than the last. Perhaps most troubling is that microplastics' presence in the world—and in our bodies—grows every year. Because, of course, their spread is linked directly to plastic production's upward trajectory. The more plastic industry makes, the more little pieces of it we all take in. One finding in Matthew Campen's brain study illustrated that dynamic vividly: The brains of people who'd died in 2024 contained nearly 50 percent more plastic than those who'd passed just eight years earlier. "This stuff is increasing in our world exponentially," Campen said. "I don't need to wait around thirty more years to find out what happens if the concentrations quadruple."

CHAPTER 7

Wilson vs. Formosa

*1997—Oceanographer Charles Moore discovers the Great Pacific Garbage Patch while sailing from Hawaii to California.
194 million metric tons of plastic produced.*

When Diane Wilson was a girl, she'd wait by the docks each evening for her father to pull in with his day's catch, watching as the sun set and the other shrimpers and fishermen returned too. The bay was the center of her world back then. It was the 1950s, and no one in sweltering Seadrift, Texas, had air-conditioning, so the salty smell would waft in through open windows, filling the house where Diane's parents were raising their seven children. Around three or four in the morning, she'd hear boat engines sputter to life. "I was so close and connected to that bay and that water," she tells me now. "To me it's alive. It has an essence."

On a map, what Wilson calls "the bay" is actually a series of bays—Lavaca Bay, Matagorda Bay, Espiritu Santo Bay, San Antonio Bay. Tucked behind barrier islands and cut through by peninsulas, these interconnected inlets of the Gulf of Mexico sit halfway between Houston and Corpus Christi on the sunbaked Texas coast. The enormous sky is a deep, sharp blue; the land flat, wide open, and nearly empty. Cotton bolls dot fields and cattle loll beneath live oaks. At the end of a bumpy road, just past a Confederate flag in someone's front yard, I wind down an unpaved driveway, through high grass, until I spot the red pickup truck and old kayak Wilson told me to look out for. Just beyond them is a little purple house with red trim. In flip-flops, a patterned skirt, and baggy tank top, she stands out front, waving to me while talking animatedly on the phone. A moment later, we're settling in under a tree with pale green Spanish moss hanging from its twisting branches, and Wilson, rocking on an old wooden swing, seems to start our conversation midstream.

The bay and her relentless efforts to protect it have shaped her life, she tells me. When she was young, its waters were rich with dolphins and oysters, seahorses and seagrass, turtles and crabs. At eight, she started helping out her dad as a summertime deckhand. By her late twenties, she was captaining her own shrimp boat. She loved the quiet when she was alone on the water, and the sense of belonging that came from being part of a fishing community. But soon, the catch began to diminish. Dolphins were washing up on shore, birds dying, toxic algae blooming. No one—at least among the fishermen—really knew why.

In 1989, Wilson was forty, raising five kids, and when she wasn't out on her boat, she was running her brother's fish house—fueling other shrimpers' vessels, getting their ice, mending nets. One day, she was nursing a lukewarm cup of coffee, rubber-booted feet on her desk, when a shrimper she knew came in with a newspaper. "He had about three different types of cancer, he had these lumps," she tells me, touching her neck. "He pitched this article at me. And he said, 'What do you think?'" The story shocked her. The Environmental Protection Agency had published its first pollution inventory, and Calhoun County was ranked worst in the nation for toxic chemicals disposed on land. "How can a county that small, and you never hear a word—there has never been any mention of pollution or anything—how can you be number one in the nation?" Wilson wondered.

WILSON VS. FORMOSA

It bothered her so much, she called a meeting to discuss it, requesting a room at Seadrift's tiny city hall. Within a couple of days, the town secretary showed up at the fish house. "And you don't get people down the fish house. You get shrimpers, truck drivers, but you don't get people dressed up," Wilson tells me with a laugh. "And she came down there in this flowery dress." The secretary told her she'd have to move the meeting out of town. "I'm like, 'Why?' And she said, 'It's puttin' up red flags.' 'Red flags'—that's all she would say." The next day, things got even stranger. Wilson had never spoken to the local bank president, but there he was at the fish house "in a three-piece suit, black shiny shoes," she tells me. "And he said, 'Diane, are you fixin' to start a vigilante group to roast industry alive?'" Before long, her brothers were getting calls demanding they stop her. They ignored the pleas, she recalls, "but they were curious. They said, 'What are you doing that they're so upset?' I said, 'I didn't do a thing. All I did was ask for a meeting.'"

That, it seemed, was enough to stir things up in a region whose economy and politics are deeply intertwined with polluting industries. Union Carbide, the company whose 1984 chemical leak in Bhopal, India, was one of the world's worst environmental disasters, had a big plant nearby. So did DuPont.

Despite the pushback, Wilson went ahead with her meeting. A day before it took place—in the end, she'd secured the elementary school cafeteria—she got an unusual piece of mail. It was a clipping from a newspaper's public notice section, accompanied by an unsigned note that simply said "Ms. Wilson, are you aware of this?" In dry, legal language, the notice reported that a company called Formosa Plastics was applying for environmental permits. "I didn't even know who Formosa was," she recalls. That was about to change. And in the more than thirty years since, "it's been like an onion. You peel one layer, and then you just find out more and more."

PLACE INC.

THE STORY OF WANG YOUNG-CHING'S JOURNEY FROM CHILDHOOD AS THE son of a Taiwanese tea farmer too poor to buy him shoes to the pinnacle of corporate success has been so carefully buffed it feels more like mythology than biography. Widely known as Y. C. Wang, he's said to have fetched water for his family each morning, fed the pigs in the evening, and walked three hours back and forth to school every day. By age fifteen, in the early 1930s, his formal education had ended, and he left his family to work for a rice seller in the island's south. A year later, with money his father raised from neighbors, Wang opened his own rice stall, and his younger brother, Y. T. Wang, soon joined him—the beginning of a lifelong business partnership. Y. C. Wang recalled later that he bathed with cold water so he'd have a few extra pennies to put back into the business. While the shop next door closed at 6 p.m., he kept his open until 10:30. Thriftiness and hard work are the central themes of his legend. His rise would parallel Taiwan's evolution from largely agricultural society to global manufacturing powerhouse, making him "a symbol of the island's economic miracle," as one writer observed.

After World War II, the brothers moved into the timber business, and by 1954, Y. C. Wang had amassed enough capital to get into yet another industry. He put $500,000 into a factory that would produce polyvinyl chloride, or PVC, powder, a raw form of one of the world's most widely used plastics. The US government, seeking to build Taiwan into a capitalist counterweight to mainland China, provided a $680,000 loan. Originally known as Fu-mao Plastics Corporation, the Wangs' new company soon became Formosa Plastics. In the early years, oxcarts carried its resins to port.

Plenty of cheap PVC powder was coming into Taiwan from Japan, but Y. C. Wang was determined not to let the competition crush him.

He decided to increase production tenfold to push costs down. "It was like riding a tiger," he said later. "You can't get off. You must go forward or get eaten." He founded another company, Nan Ya Plastics Processing, which bought Formosa's powder and turned it into pipes, plastic film, and imitation leather. Then he started New Eastern Corporation to use Nan Ya's material for products like handbags, raincoats, and shoes. Eventually Formosa Chemicals and Fiber Corporation would produce synthetic fabrics such as nylon and polyester, which Nan Ya used as backing for its imitation leather.

Expanding vertically in this way—building their own supply chain—turned out to be a profitable strategy, so the Wangs pushed it in the other direction too. When a shortage of ethylene, a key ingredient in PVC powder, slowed production, the brothers built a petrochemical plant to make it from the oil derivative naphtha. And they refined oil to get the naphtha. As the years passed, and a new age of global consumerism drove demand for plastic ever higher, their empire grew geographically too, with operations in mainland China, the United States, Vietnam, and Indonesia. The Formosa Plastics Group, the corporate home to all those ventures, eventually comprised more than fifty subsidiaries—producing not only plastic but a long list of goods including steel, textiles, electronic components, and gasoline.

Through it all, Y. C. Wang kept his drive, and his austere habits. His dark blue suits were perfectly ironed, and he had a "straight back, gaunt face, slightly protruding ears and alert eyes," a writer who knew him recalled, "with eyebrows always knitted together and a thin, pinched mouth." For decades, he rose at 3:30 a.m. to exercise, and even after he'd become a billionaire, his wife had to sneak new shirts into his closet because he resisted spending money on clothes. And he drove his employees as hard as he pushed himself. "Wang was harsh on his managers, because he wanted them to be as tough as roaring tigers on their subordinates," the journalist wrote.

PLASTIC INC.

When he died in 2008, at ninety-one, Wang was Taiwan's second-richest person, worth about $6.8 billion, and known in the country as "the god of management." Formosa had become the world's tenth-largest chemical company. Today, it's the ninth biggest, with $31 billion in annual chemical sales.

There were darker currents beneath that relentless growth. Y. C. Wang's personal life was tangled—he had ten children with three different women, while married to a fourth. He left no will, and his offspring feuded in court for years after his death. Evidence emerged that the Wangs had parked billions of dollars in overseas trusts to avoid taxes in Taiwan. There were accusations that Formosa tried to intimidate its critics: It filed a criminal defamation complaint and a million-dollar lawsuit against a Taiwanese environmental engineer in 2012 over evidence he'd presented of elevated cancer risks near one of its plants. A court dismissed the case, but the scientist said it succeeded in chilling criticism of powerful corporations.

Concerns about toxic pollution were a constant. In 1998, mercury from one of Formosa's plants, baked into three thousand tons of concrete and packed in plastic bags, was dumped in a field in Cambodia. Villagers who scavenged the unmarked site grew ill, and a cargo worker who'd handled the waste died. In 2016, chemicals spilling from a steel plant the conglomerate was building caused a massive fish die-off along two hundred kilometers of Vietnamese coastline. The prime minister called it "the most serious environmental incident Vietnam has faced," and hundreds joined street protests, a rare show of public anger in a nation that tightly controls political expression. Formosa eventually offered $500 million in compensation and cleanup.

The company was set up as an interlocking network of subsidiaries so complicated it could be hard to tell who was responsible when something went wrong. That may have been intentional. "The structure of

Formosa Plastics Group is a study in obfuscation," one advocacy group concluded. "Such complexity can deflect scrutiny and facilitates impunity for violations of regulations."

At home in Taiwan, rapid industrialization had created terrible air and water pollution, and by the 1980s a nascent environmental movement was gaining strength. Tired of the headaches activists were giving him, Y. C. Wang was increasingly looking overseas. In 1983, the company had opened a small PVC plant in Point Comfort, Texas, across a long, low causeway from the county seat of Port Lavaca, about forty kilometers from Diane Wilson's home in Seadrift. By 1988, Wang had a much bigger project in mind—a $3.2 billion complex that would employ 2,700 people to make a wider range of plastics. He thought Point Comfort might be a good place to put it. The fishing, farming, and oil industries were all hurting in Calhoun County, and unemployment had recently topped 15 percent. As far as local officials were concerned, any dangers the plant might pose paled in comparison to that threat. "Once we get our economy stabilized, then we can afford to be a little more choosy with who and what we let in," said Doug Lynch, the county's economic development chief.

Anxious to beat out Louisiana's bid for the project, Texas courted Formosa. A delegation visited Wang in Taipei, and when he reciprocated, Port Lavaca feted him with lunch at a local businessman's home—complete with roses for his female relatives and a high school choir performance. But Wang changed course, deciding to put the plant in Taiwan after all. He'd build a smaller but still sizable facility, costing $1.3 billion, in the United States. Calhoun County's representatives again flew to Taiwan. "How does a small, struggling Texas county get a brand-new, billion-dollar petrochemical plant?" *Texas Monthly* asked. "By giving one of the world's richest men everything he wants." What Wang wanted were tax breaks, expedited environmental permits, and public funding to dredge the Matagorda Ship Channel. Texas said yes to it all, and the governor and US Senator Phil Gramm stood beside Y. C. Wang as the project was announced.

Fed by pipelines delivering ingredients such as ethane, propane, and naphtha from oil and gas fields in Texas and beyond, the complex wedged between Lavaca Bay and Cox Creek began to grow. By 2021, with a fourth major expansion—costing $5 billion—complete, the original plant's two units had become twenty. Three years later, the company was already laying the groundwork for another big growth spurt. Today, about 2,500 Formosa employees and 1,000 contract hires work on a site almost three kilometers long, and nearly as wide. As I drove past (the company did not allow me inside), I craned my neck at the vast city of enormous white tanks, hulking towers, and winding pipes. It seemed to go on and on, one huge metal structure after another, flames licking up here and there from tall pipes.

Inside those structures, a dizzyingly complex series of multisyllabic chemical processes take place—ethylene fractionation, propane dehydrogenation, cracked gas compression, caustic oxidation, demethanization, among many others. There are two gas-fired power plants on-site, generating enough energy to heat and power more than four hundred thousand homes—that is, if it weren't being used for plastic production. Temperatures hotter than 800 degrees Celsius "crack" molecules of the oil- and gas-derived ingredients into the petrochemicals ethylene and propylene. Those in turn are used to make millions of tons a year of three of the world's most common plastics: polyethylene, polypropylene, and PVC vinyl. Produced in a variety of grades and densities, ready to be turned into "basically anything that you can think of that's plastic," as one company executive put it, they are loaded onto trucks and trains bound for the distributors and manufacturers who are the plant's customers. PVC is shipped as a snowy powder. Polyethylene and polypropylene are sold in tiny pellets, each about the size of a lentil. They're known as nurdles.

WILSON VS. FORMOSA

DIANE WILSON AND Y. C. WANG MET ONCE, ACROSS A TABLE IN A LOW, blocky building near the plant. He sat quietly as she detailed her objections to the expansion he was planning, and invited her to tour some of his company's other facilities. She declined. The article she'd read at the fish house in 1989 had turned her into something she'd never imagined becoming—an environmental activist. In the years that followed, she went on more than a dozen hunger strikes, tried to sink her boat atop one of Formosa's discharge pipes, served repeated jail stints, went to Bhopal, Washington, and Taiwan. Once, she pledged fifty dollars to attend a Republican fundraiser, smuggling a protest banner into a swanky banquet hall by wearing it as a shawl. When then–Vice President Dick Cheney took the stage, she got as close as she could, then unfurled the banner while shouting, "Corporate greed kills."

It took its toll. Calhoun County is deeply conservative, and most there saw the industry she was fighting as their economic lifeblood. It was all too much for her husband, and when they split, she became a single mother of five. "I've lost my house, I lost the boat, I lost the marriage, people fired me from jobs," she tells me. For a long time, she doubted herself. She hadn't gone to college, "and people were always saying 'You're the wrong person'" to take on such powerful foes. That made sense to Wilson. "I thought, 'There's bound to be someone much better than me to do this.'" Still, it felt right. Eventually, she understood why. "I realized I was the best person because I had a passion for it," she says. "I will not give up on that bay."

More than thirty years later, she still hasn't. As Formosa's plant grew, the company became her primary adversary, the target of her first hunger strike and many subsequent protests. Because she was its most outspoken—and often only—local opponent, workers with stories to share about what was happening inside its plant would often come to her. Many feared reprisals if they spoke publicly, so they'd approach her on the sly, phoning out of the blue and braving the rattlesnakes in her yard to meet at her trailer outside town.

Few were as anxious as Dale Jurasek. I meet him on a hot, still day at a marina with a view of the complex where he worked for twenty years. He tucks his rangy frame around a metal picnic table, his long face weather-beaten and lined, brown hair pulled back in a short ponytail. A chest tattoo peeks out at the top of his shirt, and another one, of a serpent, winds around a leathery forearm. Between sips from a "Don't Tread on Me" mug, he speaks slowly, his voice deep and gravelly.

He was just eighteen when he started on a construction crew building the first Formosa units, and he got hired when the plant opened. In his telling, employees handled dangerous chemicals with stunning carelessness, and routinely falsified paperwork to cover up hazardous conditions. Toxic substances would pool on the floor, workers trudged through tainted wastewater with little protective gear, and they sometimes resorted to using foam earplugs to fill holes in pipes carrying corrosive gases, he says. Liquid flowing into one storm drain "was so potent that it just eventually cut its own path through the concrete. And if you stepped in it, it'd feel like the bottom of your boots were on fire," he tells me. When it splashed, "it'd eat holes in our jeans." Ethylene dichloride and vinyl chloride—suspected and known carcinogens, respectively, and both key ingredients in some plastics—used to shoot from a pipe across the surface of one basin, he says, shaking his head and blowing through pursed lips. The chemicals are volatile, so he often feared explosions. "It'd scare you to death."

Managers treated the air and water around the plant with the same cavalier attitude they brought to workers' safety, Jurasek tells me, pulling a cigarette from a red pack and tapping it on the table beside his lighter and beat-up baseball cap. The potent substances cutting channels through concrete were running untreated off Formosa's grounds, he says.

He remembered seeing such runoff draining into a ditch that led to a nearby highway rest stop. "I didn't say nothing," he tells me with a sad, crooked smile. "We learned to keep our mouths shut."

Formosa, in Jurasek's experience, cared about little besides maximizing output—and therefore its profit. "Your operation rule is 'Run everything balls to the wall,'" he tells me. "You push everything as hard as you can." Managers had little time to worry about complying with pollution rules. Instead, they'd flush equipment clean before taking samples for regulatory agencies, dilute or switch out batches likely to violate contamination limits, and even bend needles on gauges to manipulate readings, he says.

It wasn't just chemicals running into the bay. Inside Formosa's complex, Jurasek says, the little plastic pellets known as nurdles were everywhere—spilling as they were moved from place to place, then getting swept into drainage ditches whenever it rained. Out with his family one day, he noticed pellets in his boat and realized his kids were tracking them in, their feet plastered with nurdles as they played in the water. He told supervisors, he recalls now, but no one seemed to care.

Eventually, Jurasek reached a breaking point. "I thought to myself, 'Y'know, this ain't right,'" he tells me. He contacted the Environmental Protection Agency and was soon providing information to investigators. Eventually, managers realized he was talking to authorities. He wasn't fired, but he says his bosses gave him the dirtiest, most miserable jobs. He was having health problems too—memory loss, nerve damage in his legs and hands, oozing lumps on his neck and torso.

Part of me doesn't want to believe any American plant's operations could be as egregiously reckless as Jurasek describes Formosa's, and, to be fair, he hasn't been inside the facility since 2001. Formosa, which did not reply to my requests for comment on his allegations about its handling of chemicals, says it is committed to protecting the environment and guarding workers' health and safety. But while it's hard to verify the specifics, his accusations are well within the realm of credibility. In 2005, an explosion at the plant injured sixteen workers, sending thick plumes of black smoke into the sky. A fire burned for five days, and federal investigators blamed failures in the plant's design for turning a small accident into a major disaster. The inquiry Jurasek aided never went anywhere (he ultimately settled with the company), but the Environmental Protection Agency continued scrutinizing Formosa's operations, and in 2009 the company agreed to spend $13 million to address what the Department of Justice called "serious violations" of air, water, and toxic chemical rules at the plant and another Formosa facility in Baton Rouge, Louisiana. As recently as 2021, the company agreed to a $2.85 million settlement with the federal government over fires, explosions, and toxic releases that left employees at the Point Comfort plant hospitalized with chlorine inhalation and second- and third-degree burns.

By 2012, Jurasek tells me, he'd settled into an early retirement, but he was hearing troubling things from old colleagues, and he couldn't put his time at the plant in the past. "People were calling me, telling me they were sick, that they had knots behind their ears, cancer—I didn't know what to do." That's how he ended up in a bar with a concrete floor, talking to Diane Wilson. His cooperation with authorities had made him a pariah, and his family had suffered threats and intimidation, so when he asked her if they could talk, he chose a place sixty-five kilometers away. That was nothing new to Wilson. "I've met a lot of workers in crazy places—parking lots, dark alleys," she told me. Jurasek was so afraid he wondered if even she, Formosa's most visible local critic, might somehow be in league with the company. So he gave his gun to the bartender, a friend. "If something comes down, you know what to do," he recalls instructing the guy. "He said, 'I got you, Dale.'"

Jurasek sat at the back, in a cowboy hat, and when Wilson arrived, he demanded she empty her bag and pull up her shirt to prove she wasn't carrying a recorder. That settled, he told her what he'd seen at the plant.

Much of the conversation focused on workers' suffering, but Jurasek also talked about the pellets spilling into the water. Wilson had seen nurdles around, but until then hadn't given them much thought, since her concerns had centered mostly on toxic chemicals.

Around the same time, she heard that Texas RioGrande Legal Aid, a pro bono lawyers' group, was expanding its coverage area to include Calhoun County. "I went to the first meeting, and I said, 'Do y'all do environmental stuff?'" Wilson told me. They did, and she soon began to work with them, looking for ways to push Formosa to pollute less. As she grew increasingly concerned about nurdles, she filed complaint after complaint with state regulators. They visited the site, but little came of it, and eventually Wilson realized no one in power was going to stop plastic from pouring into the bay she loved. To end the spills, her lawyers told her, she would need to collect evidence to prove just how egregious they were. In January 2016, that's exactly what she set out to do.

DESPITE THE SHADE ON HER PATIO, THE SWEAT IS DRIPPING DOWN MY NECK and back as Diane Wilson explains how she and a handful of other volunteers started gathering nurdle samples, working under the auspices of a group she led called San Antonio Bay Estuarine Waterkeeper. She's moved from the bench swing to a couch covered with a worn quilt of brightly colored squares; behind her, vintage license plates are tacked onto the side of a crooked treehouse that wraps around one of the big oak's thick branches. Dale Jurasek and his son sometimes helped her comb the water's edge, but mostly it was Wilson and another former Formosa worker, Ronnie Hamrick, who'd head out every day to collect samples.

Hamrick had spent twenty-five years working for Formosa, and he had stored up plenty of anger toward the company by the time he took retirement. "I falsified a lot of reports" to cover up reckless handling of dangerous chemicals, he told me. "It got to the point where I couldn't stand it, the lying." So when the opportunity came to do something that might rein his old employer in, he grabbed it. "You're gonna have to have a lot of samples to beat this company," he remembered telling Wilson. "Otherwise, you ain't got a chance."

Wilson agreed. "We started wading out on the bay," she recalls, "along the shores, around the boat ramps," poking through high, marshy grasses and stretches of pebbly sand. She soon saw the problem was much bigger than she'd realized. Once she knew where to look for nurdles, "they were everywhere." She and Hamrick would scoop them up with fish nets, carefully noting the time and location of each sample, and taking pictures and videos. Sometimes they'd get enough at one spot to fill a four-liter bag, other times they needed a big trash sack. By a boat launch on Cox Creek—wide and muddy, meandering right past the plant—white pellets regularly carpeted the marshy shore, "like that deep," Wilson recalls, holding her hands about twelve centimeters apart. They found PVC powder too—raw vinyl—sparkling in the water and etching lines along the shore.

The work was painstaking. Hamrick would sometimes be out for eight hours at a stretch, in the blazing sun and the rain, and when temperatures dipped down near freezing. No matter the weather, "we had to get samples," he told me. "It's a full-time job, that's the bottom line." About six months in, Wilson bought a cheap kayak and started paddling up and down the creek, a couple of hours each way. She scooped pellets from the water as she went, but she was really hunting for something bigger: Formosa's discharge points, known as outfalls. "I didn't have a map to show where they were"—neither the company nor the state environment agency would share that. The first one she found was fed by a ditch "coming right from the plant. And it was going right to the fence, and it was going right into the creek," she says.

The samples piled up, in bins and boxes stacked in Wilson's barn.

"Some were trash can–fulls, some were garbage bag–fulls, some were baggie-fulls," each representing a particular visit to a specific location, she told me. There were bottles of powder-laden water too. Every sample, she knew, was bringing her one step closer to facing down Formosa in court.

WILSON'S LEGAL TEAM FILED SUIT AGAINST THE COMPANY IN JULY 2017. AS the case moved slowly forward, they began getting glimpses of what was going on inside the plant. Formosa had to release documents and emails, and allow her engineering expert to examine its facility. All the while, she and the other volunteers were still gathering samples daily. Finally, in March of 2019, Diane Wilson hooked a trailer onto her pickup truck, piled it with 2,400 of those samples—46 million individual nurdles, she estimates—and drove to federal court. "We started carrying them into the courthouse," one of her lawyers, Amy Johnson, told me. Security guards at the entrance were befuddled, so "they called the judge and said, 'There's all this stuff, do you want it?'" He came down to take a look and, after peering into the bins, told Wilson he'd need only a few in the courtroom.

Formosa's central defense was that any plastic releases had been merely "trace" amounts, and therefore allowed under its permit. The voluminous evidence Wilson and the others had gathered—hundreds of photos and videos, along with the pellet and powder samples—was key to rebutting that argument. While questioning one of the company's expert witnesses, "he was talking about 'trace,' and I had a video of Diane in a kayak on the creek," Johnson told me. "And all around her is a bed of plastics floating on the water, probably five or ten feet [one and a half meters to three meters] out," she said. "We all know that's not a trace amount."

Others also testified how plain the problem was to see in the waters of Cox Creek and Lavaca Bay. Dale Jurasek's son Cheyenne said he'd found "handfuls" of pellets on the shore when he visited a clubhouse Formosa owned. "It literally looked like someone took a five gallon [twenty liter] bucket and just dumped it." Texas A&M University environmental scientist Jeremy Conkle, one of Wilson's expert witnesses, submitted a report that said the samples "demonstrate that plastic and powder were a constant presence" at the dozen sites Wilson's team had visited over more than one thousand days. He'd made his own visits too. On one such occasion, pellets "covered the ground, looking like a dusting of sleet or hail" beside Cox Creek, he'd written. Not far away, he saw PVC powder in amounts he called "shocking," deposited in a line along the shore. "It reminded me of the rings seen in dirty bathtubs," he wrote. "I have never seen anything like what I witnessed in Cox Creek and Lavaca Bay."

Formosa had hired a cleanup contractor after Wilson started sampling, and its crews collected more than eighty-five thousand bags of pellets and powder over about two years. By Conkle's calculations, that amounted to at least 170 tons and 7.5 billion individual pellets—and perhaps as much as 1,600 tons and 75 billion pellets (the wide range resulted from varying estimates of how full the bags were). "Despite the removal of this astonishing amount of plastic," he wrote, "cleanup continues, and I observed more newly released pellets and powder" in 2018. Indeed, much of Wilson's effort was aimed at showing the pellet releases were ongoing and that the debris in the creek and the bay wasn't only from past spills, as Formosa claimed. A manager from the cleanup company told lawyers that crews would regularly remove all the pellets they could see by a dock on Lavaca Bay, then find more the next time they went.

Wilson isn't easily shocked, but one number that came out at trial stunned her. The plant, a Formosa lawyer said, was making a trillion pellets a day (although a company manager later said output was more like three hundred billion daily). Either way, in Wilson's view, that scale explained why so much plastic was ending up in the water. Preventing nurdles—even large quantities of them—from spilling simply wasn't worth the company's time or trouble. Anything that stopped or slowed production would cost money. To Formosa, she believes, losing plastic is just the cost of doing business.

That assessment certainly fit with testimony from Aiza Jose-Sanchez, Wilson's engineering expert, who had examined the plant's waste and stormwater systems. She'd concluded Formosa's pellet and powder problems were systemic, a direct result of the way the plant was set up. In some parts of the compound, any water present was deemed to have been in contact with hazardous substances and required treatment before being discharged. In other areas, known as "non-contact zones," rainfall or other water could be shunted off-site untreated. Nurdles and powder, Jose-Sanchez found, were littered throughout the non-contact areas—lining drainage ditches, sitting on the ground—and were swept off to the creek and bay whenever it rained. What's more, she said, Formosa's drainage systems were too small, so they often overflowed, sending water—and whatever was in it—coursing over or around gates and screens meant to filter it. In any case, those screens were liable to clog, which meant that even without a flood, pellet- and powder-laden water flowed right past them. Van Rozner, a former employee, said PVC powder would sit in big piles workers eventually washed into creek-bound ditches. In one photo he took, inside an open-air building, it looked like there'd been a blizzard, with plastic powder piled up like snow, as high as sixty centimeters in places.

It was exactly what Dale Jurasek had described seeing when he worked at Formosa. And he testified he'd told his bosses about pellet loss in 2000. Indeed, one of the most damning realities that came to light in court was that Formosa had long known it was spilling plastic. Internal audits detailed the very failings Jose-Sanchez described. Photographs from one report done for the company showed pellets and powder in stormwater culverts, on a railroad line, beside drains, and in loading areas and production units. "The ditches are full and pellets are floating on the water," said a report from March 2016, less than two months after Wilson had started her sampling—and while Formosa was vehemently denying it had a problem. Many photos showed screens that were meant to capture nurdles ripped, missing, or left open. "During rainfall pellets will make their way over the concrete wall," one audit document warned.

There was more. In 2010, a marine biologist Formosa had contracted to monitor its discharge told two company executives she'd found pellets at one of its outfalls. "We noticed plastic material floating in the bay again that was coming from the diffuser," the scientist, Lisa Vitale, wrote in an email. "There was quite a bit of it this year, and I just wanted y'all to be aware." Her message reminded its recipients that it wasn't the first time she'd seen their nurdles in the bay. "If you remember, this has happened before, in October 2004 and again in October 2005, we previously sent y'all a sample to analyze and y'all discovered it was coming from the plant." She told Wilson's lawyers she'd notified the company of plastic discharges at least five or six other times.

Formosa's lawyers put forth a slew of arguments. Wilson and the others shouldn't be allowed to sue because they hadn't been directly harmed by the spills, the defense said. The company claimed an agreement it had reached with the state environment agency to pay $122,000 for six permit violations made Wilson's whole case moot. Beyond those efforts to brush the suit aside on procedural grounds, Formosa's main defense was that it was doing its best to minimize spills, and any discharges were within the "trace" amounts allowed under its permit. "It would literally take me all day to describe to you all the improvements that Formosa made" in the year since Wilson filed her suit, one of the company's lawyers told the judge. She detailed some of the measures in place: Workers regularly checking stormwater ditches for pellets and powder; vacuum trucks scooping up stray nurdles; filters and floating booms to stop them from escaping. Given such efforts, Formosa's lawyers argued, any plastic Wilson found in surrounding waters was likely old, the accumulation of more than twenty years' worth of small discharges, or else had come from other companies.

After a four-day trial, US District Judge Kenneth Hoyt issued his ruling. It was more than Wilson had dared hope for, a vindication of everything she had been saying. One by one, he rejected the company's arguments. All told, the judge determined, Wilson had proved, over three years, 736 days of illegal releases at one of Formosa's outfalls, and 1,149 at a group of eight others. "These releases are extensive, historical and repetitive," Hoyt wrote. "The violations are enormous," breaching both the federal Clean Water Act and the terms of the company's state permit. Formosa had compounded its rule-breaking by failing to inform regulators of even a single violation, which it was obliged to do. Hoyt chided the state agency too. "Based on the overwhelming evidence," the judge wrote, the $122,000 fine the Texas Commission on Environmental Quality had levied "merely shows the difficulty or inability of the TCEQ to bring Formosa into compliance." His bottom line was brutal: "The Court concludes that Formosa is a serial offender."

VICTORY, DIANE WILSON TELLS ME, "BLEW MY MIND. FOR ALL THOSE YEARS, I had never tasted justice, and I got it for the first time," she says. "It was totally, totally amazing." It wasn't over yet, though. Next, the judge would have to set the penalty for Formosa. But the company was clearly rattled, and rather than wait, executives decided to settle. When Formosa's representative walked into negotiations, "she looked at me," Wilson recalls, "and she said, 'You are one persistent woman.'" Wilson laughs. "I have fought them for thirty years. And I have been very persistent." In her memory, Formosa's starting offer was $15 million. "I said fifty million dollars." The executive "went in and talked to

New Jersey [Formosa's US office], and she came back and said, 'Thirty million dollars.'" Wilson's lawyer, Johnson, glanced over, Wilson recalls, "and I said fifty million dollars." The executive "talked to Taiwan [corporate headquarters], and came back in. And: fifty million dollars."

It was an extraordinary sum, the largest-ever settlement of a private citizen's lawsuit under the Clean Water Act. "OK, Diane, where do you want to put it?'" she recalls Johnson asking. The money would go not to her personally, but to a trust funding local environmental work. Its biggest initiative, at $20 million, is the creation of a sustainable fishing cooperative, aiming to revitalize the bays' ecosystems so small fishermen and shrimpers have a future. Other grants fund beach restoration, kids' environmental education at YMCA camps, and a slew of scientific studies on local ecosystems and microplastics in the environment.

More important than the money, Wilson says, were the commitments she extracted from Formosa to stop spilling plastic and pay to clean up its mess. The agreement the lawyers hammered out gives her an unusual degree of involvement in holding the company to those promises. Her team scrutinizes the changes Formosa must make to its plant and its cleanup plans. She can challenge any of it in court, and Formosa covers fees for her lawyers and engineering expert.

Meanwhile, an independent monitor tracks new spills. Whenever pellets or powder are found in Cox Creek or Lavaca Bay, Formosa is fined, with the money going to Wilson's trust. The size of that per-day, per-body-of-water levy climbs, from $15,000 in 2020 to more than $68,000 in 2025. Four and a half years in, Formosa had already racked up more than $25 million in penalties.

Wilson believes it all adds up to a powerful model—citizen science and activism holding a major polluter to account, and citizen enforcement making the changes stick. An industry lawyer I spoke to agreed: "The reverberations have been far-reaching," said Karen Hansen, who represents companies on water-quality issues. The costs Formosa was saddled with—not just the $50 million settlement and ongoing fines but the expense of cleaning up ($29 million for just the first phase of nurdle removal), overhauling its plant, paying Wilson's lawyers and engineer—may not be crushing for a company of its size, but they are far larger than any penalty state or federal regulators have imposed. And they've served as a sharp warning to executives elsewhere in the industry, Hansen told me. "No company wants the liability that Formosa Plastics found itself with," she said. To avoid it, they realized they'd have to address their own nurdle leaks. Companies that once might have thought their pellet discharges fell under the definition of "trace" now know a judge may see things differently.

That matters, because, of course, this problem is far wider than a single plant. Indeed, pellets are lost almost everywhere they are made, shipped, stored, and transformed into goods and packaging. More than seven hundred million spilled from a cargo vessel on the Mississippi River near New Orleans. Forty-nine tons' worth, belonging to Saudi Arabia's SABIC, were dumped when a storm smashed two shipping containers into the water in Durban, South Africa. In what is believed to be the biggest plastic spill on record, a container ship carrying some 1,700 metric tons' worth, or about seventy billion individual nurdles, sank in the Indian Ocean in 2021. Pellets washed up on hundreds of kilometers of Sri Lanka's coastline—two meters deep in places, so thick "you could not see the sand," a local environmentalist said—and were found clogging the mouths of washed-up fish and in the bodies of dead dolphins, turtles, and whales. British plants lose between five billion and fifty-three billion nurdles a year.

Governments have yet to do much about it. So while some companies have begun cracking down on spills, other activists are following Wilson's lead, stepping in where regulators have failed to act. After the Formosa verdict, South Carolina environmentalists started collecting nurdles around Charleston Harbor. They won a $1 million settlement—and an agreement to make changes to prevent future spills—from Frontier Logistics, a plastics distribution company. Other such lawsuits are likely. Wilson has spoken to audiences from Harvard to Copenhagen, and fields frequent inquiries from activists interested in replicating her approach.

While the specifics of Wilson's case are unique, she is part of a tradition of concerned Americans stepping up to fight big polluters. Still, it's disheartening to remember why such extraordinary efforts are necessary—because government hasn't done its job.

WITH ENFORCEMENT NOW THE FOCUS OF WILSON'S CASE, RONNIE HAMrick still goes out every day collecting nurdles and plastic powder. He arrives at the marina where we've agreed to meet in a pickup truck driven by another volunteer. Four big dogs tumble out of the cab with the two men, and Hamrick, in a camouflage baseball cap and faded gray T-shirt, squints in the sun. His tanned arms are mottled with patches of white, and an old yellow-green towel is stuffed in his jeans pocket. Clambering over big chunks of concrete beside a wooden pier, he bends down to sift through the twigs and dried grass that have accumulated around them. After a moment or two, he holds a dozen white pellets in his cupped palm. "You want nurdles, you just sit right here," he says. "I can just find all you want."

Later, we drive over the causeway out of Port Lavaca and park right across a two-lane highway from Formosa's vast complex. With my back to the plant, it's a beautiful spot. Fluffy clouds dot the sky, and just below us stretches of lush, swampy vegetation surround the muddy water of Cox Creek. Hamrick grabs a rake and fish net from the back of the pickup, and we step over a low metal fence and down a steep embankment to the creek, which sweeps right past the plant. "What are you looking for?" I ask as he peers around. Alligators, he answers. I glance nervously at a bump in the water, but he reassures me it's just a stick.

Hamrick reaches his rake into the thick vegetation at the creek's edge. When he pulls a clump of plants to one side, white nurdles float in the patch of water he's exposed—hundreds of them, in a little puddle maybe sixty-five centimeters square. Pressing down on vegetation brings pellets to the surface. More are lodged in the soil, visible in layers when he lifts up clods of dirt or moves a rock. They're laced densely through the root system of a plant he yanks from the soil, like tiny eggs. When the muddy water drains out of his net, hundreds of pellets remain. "I'll get you more over here, I see a bunch of them," Hamrick tells me. He's seen it much worse than this, days when a few scoops of his net would fill a big bag with nurdles, "thousands and thousands and thousands" in just a few minutes.

There's something disjointed about the way Hamrick speaks. His stories are often meandering and roundabout, and it makes sense when he tells me he's suffered central nervous system damage—he thinks it may be from chemical exposure during his years at Formosa. There's an obsessiveness to him too, especially when it comes to these little plastic pellets. In some ways, that's served him well, keeping him out here every day, despite the lurking alligators, the hungry mosquitoes, and the brutal sun, gathering the evidence that is forcing a vast global corporation to change. "Take another picture," he urges me again and again. "You want a picture of everything." When I was sitting with Wilson, her phone started beeping relentlessly, and she laughed when she saw it was Hamrick, sending photos of that day's sampling. "We can't just quit," he tells me. "If Formosa comes back and they say there's nothing in the water—well, we've got proof."

CHAPTER 8

"They Want to Crush This Bug"

*1998—Keurig introduces its first single-serve coffee-brewing pod.
203 million metric tons of plastic produced.*

Sten Gustaf Thulin always liked to tuck a folded-up plastic bag in his pocket when he went to the store. Thulin, a Swedish engineer, invented the modern plastic bag in 1959; trillions have been produced since. Today his creation, omnipresent yet ephemeral, is perhaps the ultimate embodiment of disposability. But Thulin thought of it as reusable. "To my dad, the idea that people would simply throw these away would be bizarre," his son later recalled. Through the 1970s and '80s, Raoul Thulin said, Sten brought his bags shopping, "just naturally, because—well, why wouldn't you?"

The first plastic bags, sold for garbage or sandwiches, or to wrap dry cleaning, became available in the 1950s. But coming up with a version strong enough to carry a load of groceries was trickier. The invention of high-density polyethylene, in 1953, offered new hope of cracking that challenge, and sent engineers scrambling for a workable design. Thulin and some colleagues at Celloplast, a Swedish packaging company, began segmenting long tubes of plastic film into pieces, each sealed at the bottom and left open on top. He came up on his own with the idea of adding holes to serve as handles. Strategic folds brought it all together, and the design, first patented in 1962, proved revolutionary. Industry dubbed it the "T-shirt bag"—for its shape resembling a sleeveless undershirt—and it would become one of the most ubiquitous consumer items on earth.

The bags gained popularity in Europe, and soon Celloplast introduced them in America. Mobil Chemical was already producing Hefty trash can liners and rolls of bags for produce. But it hadn't been able to come up with a shopping bag cheap or appealing enough to compete with the familiar brown paper sack, which shoppers liked because it stood straight while being packed and when stowed in a car's trunk. Mobil executives saw immediately that Thulin's design was what they had been looking for—and they believed it could grab paper's place at supermarket checkouts, Bill Seanor, who worked on bags at the company, told author Susan Freinkel. Rather than replicating the paper sack's flat bottom, Thulin had drawn "on the distinctive virtues of polyethylene to create a wholly new kind of bag," Freinkel wrote. Today, she observed, it's easy to forget "what an engineering marvel it is: a waterproof, durable, featherweight packet capable of holding more than a thousand times its weight."

In 1976, Mobil's lawyers got Celloplast's US patent overturned, and soon the chemical giant was producing its own version of the bag. But retailers and shoppers still needed nudging. Plastic had taken over much of the supermarket, but shopping bags—a $600 million market—were "the last stronghold" for paper, Mobil's marketing manager, Ronald Schmieder, told the *Los Angeles Times*. "We are going after that."

But it would take "a lot of re-educating to get people to accept plastic," Celloplast's top US executive said. So the industry set up the Plastic Grocery Sack Council, which launched a publicity campaign, and began training store managers to push plastic on shoppers and showing cashiers how to pack items in the new bags. When Mobil's salesmen met with retailers, they'd pile four six-packs of soda into a bag and hang it on a rod. Two hours later, grocers could see it was still intact.

That marketing push helped, but the real breakthrough came when a new polyethylene iteration made the bags stronger, thinner, and cheaper. The earlier ones "couldn't compete—they were like the Gucci bag, they were so expensive," Mobil's Schmieder said. Now that had changed. In 1982, two of the country's biggest supermarket chains, Safeway and Kroger, began offering plastic bags. After that, "it was pretty much over," a longtime industry executive said. "The feeling within the plastics industry was 'this is the future—plastic is going to dominate the landscape.'" Plastic went from 5 percent of the US bag market to 40 percent in six years, and its share would only climb. "Grocer resistance to the new product melted in the face of enormous savings," the *LA Times* explained. Mobil made more than four billion bags in 1988, and sales, one of its managers said, were "growing by double-digit percentages a year." By 2011, the average American, by one estimate, used five hundred bags a year—and globally a million were consumed every minute.

Today, an almost unfathomable five trillion plastic bags are used annually, the United Nations estimates. Sten Gustaf Thulin surely would have been stunned by that spread. His now-iconic design was ingenious, to be sure, combining convenience, flexibility, and strength with low cost. But its metastasis into an astonishing volume of waste wasn't the doing of consumers toting groceries home. "Our job was and should always be to open plastics markets and keep them open," a plastic industry lawyer once observed. Companies had waged an aggressive and determined campaign to push plastic bags into our lives. While they measured their success in dollar signs, not everyone agreed the bags' proliferation was a good thing. "Does society really need plastic bags?" a prominent environmentalist asked, as they took off. Industry's next job, it was clear, would be to squelch such doubts before they turned into action.

PLASTIC INC.

AS HE FINISHED A MEAL AT MCDONALD'S ONE DAY IN THE 1980S, STEVEN Englebright noticed the pile of trash on his tray "seemed more than what I had just eaten." It got him thinking about the waste overwhelming Suffolk County, on the eastern end of New York's Long Island, where he had recently been elected to the county legislature. Trained as a geologist, he brought a scientific eye to his political work, and he'd been interested in environmental issues since his teens, when he watched development gobble up the Bayside, Queens, woodlands he'd once played in.

Englebright was especially worried about plastic. So not long after his fast-food epiphany, he drafted a bill to prohibit retailers from providing plastic shopping bags. It also proposed barring food and drink sellers from using plastic containers, including those made from polystyrene foam, or Styrofoam.

Plastic producers had gotten wind of Englebright's proposal, and their representatives showed up in force at a hearing on his bill, he tells me, his words thoughtful and deliberate, neatly combed white hair parted on one side. "I was not really imagining that national corporations would get involved with a local proposed ordinance in suburbia," he recalls. "We'd never seen anything quite like it—lobbyists being flown in" to speak to legislators. Led by the Society of the Plastics Industry, oil and chemical companies including Mobil and Amoco said discarded plastic was harmless, and claimed incinerating it reduced pollution by pushing burning temperatures high enough to eliminate toxins from other types of garbage. They warned the bill would hurt local businesses and argued that paper bags, being thicker, would increase trash volumes (they didn't consider the possibility shoppers might use fewer bags).

But timing was on Englebright's side. As his bill moved forward, the *Mobro 4000*, a barge loaded with more than three thousand tons of New York trash and nicknamed the "Gar-barge," was turned away from North Carolina, its intended destination, and wandered nearly ten thousand kilometers over five months, up and down the East Coast, and around the Caribbean and the Gulf of Mexico, unable to find a port that would take its cargo. It was front-page news, and Johnny Carson was even cracking wise about it on *The Tonight Show*. The saga focused attention on just how much trash the nation was generating, and despite the lobbying, Suffolk lawmakers approved Englebright's bill unanimously in 1988. The county executive lauded it as "a creative and innovative approach" to reducing waste, which, because runoff from local landfills was contaminating drinking water, was already getting shipped off Long Island. "Maybe we won't have to truck McDonald's boxes to Pennsylvania or put them on a barge." Industry was furious. "All this bill does is discriminate against plastic products," a Mobil spokesman said. "It doesn't offer any solutions."

Suffolk County's law was the country's first-ever ban on plastic bags and containers. By now, Englebright understood why industry's opposition was so vociferous. "They want to crush this bug, right here where it first started," he said at the time. "It's pretty clear that they had big plans for making a lot of plastic," he elaborates now, and Suffolk's restrictions posed "a threat to that business model," since other local governments might follow its lead. Plastic makers agreed the fight was bigger than one county. "Several years from now we may look back on 1988 as the opening round in a solid waste/packaging war," an industry lawyer wrote.

The Society of the Plastics Industry had lost the first battle, but it was determined to win that war. The group sued Suffolk, which it said should have conducted an environmental impact study before enacting the ban, to assess the effects of pushing shoppers toward paper bags. Not long before the law was to come into force, a judge agreed, putting Englebright's measure on hold until the review was completed. "I am going to lift up my plastic-foam cup and salute the judges," a local business coalition leader said when an appeals court affirmed the decision.

Eventually, in 1991, New York's highest court allowed the ban to come into force. But the lobbyists weren't done. Industry pleaded for time to build recycling infrastructure and prove plastic could be used sustainably. In response to the Suffolk measure, Amoco—then a big oil name, later absorbed by BP—had begun a Styrofoam recycling trial. Machinery "separates the plastic from hamburger and french fry scraps, then cleans and shreds it," *The New York Times* explained. The pellets that resulted could, at least in theory, become food trays, trash bins, and more. If the trial was successful, Amoco said, Styrofoam recycling could be rolled out more widely. "The markets are there and waiting," the plant's director proclaimed.

"It was all baloney," Englebright tells me. "It was nonsense." But the promises landed on receptive ears in Long Island. By now, a recession was biting, and Democrats had lost their legislative majority in Suffolk County. Plus, the law's most relentless champion was no longer around to defend it. Steven Englebright had gotten a promotion, winning election to the state assembly, in Albany. His old colleagues put the plastics ban on hold for a year. "The body is in the ground," one said. "Now all we have to do is carve 'Rest in Peace' on the headstone."

Soon they would. Suffolk County repealed the law for good in 1994, replacing it with a measure encouraging recycling. Between the lawsuit and the repeated postponements, the ban had never actually been enforced.

DESPITE THAT FAILURE, LOCAL GOVERNMENTS ELSEWHERE IN THE COUNtry began considering their own restrictions on bags and other single-use plastics. California was particularly fertile ground. Famously liberal Berkeley was the first in the country to ban Styrofoam food containers, getting in ahead of Suffolk with a pair of laws passed in 1987 and 1988. But in the years that followed, waste worries seemed to fade along with memories of the wandering "Gar-barge." Industry's two-pronged strategy—talking up recycling while running a gauzy, $250 million ad campaign touting plastic's usefulness in everything from bike helmets to bulletproof vests—succeeded in damping down what had looked, for a moment, like a rising tide of new plastic restrictions.

Within a few years, such bills were back—this time focused even more specifically on bags. Though bags were a tiny fraction of the waste stream, their growing presence in daily life—handed out by the dozens in supermarkets, stuffed into cupboards, tangled in trees—made them, for many, the most visible representatives of a wider problem. Along with that symbolic status, they wrought real harm. Blown into waterways and sometimes carried to the seas, bags were despoiling beaches and harming wildlife. They accounted for nearly a fifth of the waste clogging Southern California's storm drains, which, over fifteen years, the region's cities spent $1.7 billion clearing. People often tossed bags, incorrectly, into recycling bins, and San Francisco had to stop its recycling machinery at least twice daily so workers with box cutters could untangle them from conveyor belts—at a cost of $700,000 annually. Dealing with littered bags sucked another $8 million from city coffers. Disposable bags may be free, author Freinkel observed, "but that doesn't mean they have no cost."

By 2004, San Francisco was considering requiring stores to charge for single-use bags—not just plastic ones but paper too. That approach would have solved the (real) problem industry had seen in Steven Englebright's bill—the danger that restricting plastic bags would simply push people toward paper, bringing a different, but not insignificant, set of environmental harms. By turning bag use from a default to a decision, charges—typically between five and twenty cents—reduce consumption while allowing flexibility for those who want a bag, and many advocates prefer such fees to outright bans.

But in 2006, under pressure from plastic bag makers who slammed the idea as "a tax [that] is going to hurt those who can least afford it," the state barred local governments from requiring fees for them. So instead, San Francisco in 2007 became the first American city to ban plastic bags—nearly twenty years after Steven Englebright's doomed bill passed on the other side of the country. "Right before our eyes we see habits changing for the better," a city supervisor said when it took effect. The law imposed no restrictions on paper bags, and their use soon quadrupled. San Francisco would later correct that mistake, and elsewhere subsequent laws often restricted all disposable sacks, not just plastic.

As industry had feared, the ban proved contagious. Once again, municipalities across the country began considering bag laws. This time, the counteroffensive was even fiercer. The American Chemistry Council—representing petrochemical giants like ExxonMobil, Chevron, Dow, and DuPont—led the charge, creating an offshoot group for bag manufacturers, the Progressive Bag Affiliates (later the American Progressive Bag Alliance). The Society of the Plastics Industry, whose members included producers of plastic products and packaging, joined the fight too.

Soon, lobbyists "were going to every single city council meeting and board of supervisor meetings," one opponent later recalled. In Laredo, Texas, where bags were piling up beside the Rio Grande, the city council held dozens of hearings on a 2008 proposal to ban them. Lawmakers abandoned the idea after an American Chemistry Council rep showed up and "lobbied the City Council to help kill it," an environmentalist there said. When Seattle's city council mandated a twenty-cent fee on plastic bags the same year, the Chemistry Council helped bankroll a signature drive to get it put to a public vote, then underwrote a $1.4 million ad push urging residents to scrap the charge. The fee lost in a landslide and was quickly repealed.

Washington, DC, had better luck. It imposed a five-cent fee on both plastic and paper bags in 2009, directing the proceeds toward cleanup and protection of the Anacostia River. Three years after the law took effect, 80 percent of residents said they carried reusable bags, and disposable bag use had dropped by 60 percent.

And in California, dozens of cities and counties were enacting bans. After reading that some widely held ideas about bags' harm were misconstrued, a Bay Area lawyer named Stephen Joseph started the Save the Plastic Bag Coalition to defend them. In his view, they'd "been subject to a campaign of myths, misinformation and exaggeration." Bag makers hired him, and the industry eventually sued, or threatened to sue, nearly every California locality that prohibited bags. Most often, Joseph deployed the same argument that had initially stopped Steven Englebright's law. Before banning bags, he contended, municipalities had to do environmental impact assessments—sometimes costing hundreds of thousands of dollars—to evaluate the harms of increased paper bag use. If cities tried to avoid that problem by mandating fees for paper bags, he accused them of violating a requirement that taxes be approved by a supermajority in the state legislature.

Time magazine christened Joseph "the Patron Saint of Plastic Bags," but—at least in California—he would not prevail. In 2011, the state's supreme court rejected his arguments, and 150 Californian municipalities eventually enacted bag bans (most also required fees for paper). As was true elsewhere, the laws typically applied only to bags provided at checkout, not the thinner ones used inside stores for produce or meat.

Seeking to snuff out the threat of a statewide ban, the American Chemistry Council had plowed $6 million into California lobbying in 2007 and '08. Another such push got underway in 2010, as the legislature again considered a ban. The Chemistry Council and Hilex Poly, one of the country's largest bag makers, spent $2 million in just three months on campaign contributions and an ad blitz—the biggest lobbying spend in the state for the period. "California is in trouble," one ad's narrator intoned, ticking through statistics on unemployment and the deficit.

"What are some Sacramento politicians focused on? Grocery bags." On a website called Stop the Bag Police, the Chemistry Council warned that, if passed, the bill would kill one thousand manufacturing jobs and inflict a billion-dollar "hidden tax" on shoppers. "I've never witnessed this kind of opposition," said the bill's sponsor.

The state senate rejected the measure, but the idea kept coming back. In 2014, California became the first state to ban single-use plastic bags, letting stores provide paper or reusable plastic ones for ten cents. Undaunted, the American Progressive Bag Alliance hired consultants and a petition management company—price tag $3 million—to gather the signatures needed to call a public referendum on the matter. Once it was pending, the law—yet to come into force—was put on hold. Industry spent millions more trying to sway voters, but in the end Californians approved the ban.

Still, delay had been lucrative. While the law was on ice, manufacturers sold more than $150 million in bags in the state, the *San Francisco Chronicle* estimated. "It's a pretty cost-effective investment," a ban advocate noted. "Even if they lose the election, they make money." Industry had scored another win too. At lobbyists' urging, the state bag ban allowed retailers to sell reusable plastic bags for ten cents. A new type of bag—thicker than the old ones but similar in appearance—became common, and retailers and customers often treated them as disposable. Other states and cities restricting bags allowed the same loophole, and a decade after the ban was first approved, California lawmakers finally fixed that mistake, banning the heavier bags too.

Even with the flaw, it was clear when it came into force that California's ban was part of something much larger. Indeed, America was lagging behind the rest of the world. A flurry of bag restrictions passed in the early years of the new millennium, from Bangladesh—where they were blocking drains and causing floods—to Kenya, Uganda, and Eritrea (although enforcement and effectiveness sometimes proved patchy). In Ireland, bag use dropped by more than 90 percent after a 2002 law imposed a fifteen-cent charge on them. China barred shops from providing free bags in 2008, and their numbers fell by two-thirds. Britain was a bit later, requiring a five-pence fee for single-use plastic bags in 2015, then doubling it six years later. Data from big supermarket chains show the number of bags distributed has fallen by 97 percent, with the average shopper now acquiring just three annually, compared to 140 before the charge took effect.

Despite industry's best efforts, bag restrictions were spreading in the United States too. It wasn't an accident that they often bubbled up from the local level. Local politicians are typically more responsive than higher-ranking ones to citizens' priorities. And powerful industries can mobilize more effectively in state capitals and Washington, where established lobbyists leverage long-standing relationships, and money sometimes matters more than public opinion. So as the 2010s wore on, municipalities from Portland, Maine, to Portland, Oregon, passed their own bag restrictions. As one plastic bag lobbyist observed, "We can't visit every little town in the United States." To stop such laws from proliferating even more widely, industry would have to raise its game.

BISBEE, ARIZONA, LEANS HARD INTO ITS QUIRKINESS. A HIGGLEDY-PIGGLEDY town built into the sides of a steep gulch, it became a magnet in the 1970s for artists and hippies drawn to its dramatic beauty, chilly-for-Arizona weather, and homes whose prices had plummeted when the copper mine that long anchored the local economy closed. Twelve kilometers from the Mexican border and 1,600 meters high, Bisbee is nestled in the red, brush-covered Mule Mountains, a half-hour climb through cacti and craggy hills from its more famous neighbor, Tombstone. These days, the town's lovingly cultivated retro vibe and oddball take on the Old West have made it a popular destination for road-trippers. Vintage cars are everywhere—I saw old New York Checker cabs, an Edsel, even a hulking Greyhound bus—some shiny and restored, others rusting and dilapidated. On the edge of town, a lookout point teeters on the rim of the old copper pit. Even through a fence, it's dizzying to peer hundreds of meters down to the reddish-brown water pooled at the bottom.

David Smith was never a hippie. He'd been an arson homicide detective in Tucson, but he fell for the high desert on family vacations, and after their kids grew up, he and his wife moved to Bisbee. Sunglasses pushed up on his head and graying hair brushed back, in a boldly patterned short-sleeved shirt, he is warm and garrulous when we meet for coffee on a street where shops selling tie-dye T-shirts, cowboy boots, and Native American jewelry have replaced the saloons that once filled the place.

Smith was still a Republican when he got elected councilman in the overwhelmingly Democratic town. But later, when a group of residents suggested barring stores from providing plastic bags, he thought it sounded like a good idea. "They were hanging on all of our cactus and all of the brush," he tells me. Locals called them "desert flags," and the sight was "flat ugly. It was trashy. You have beautiful mountains and so forth, and they're covered up in plastic." The proposed ban had widespread support. One woman showed up at a hearing with "literally hundreds of plastic bags that were all sewn together, that she was wearing," he recalls, chuckling. "That's Bisbee." It was 2012, and the council chamber, and a patio outside, were packed for the vote. Normally, "you're lucky to have a dozen people" at a meeting, but this time, "there were probably a hundred and fifty," Smith says.

There was some opposition from residents who saw the five-cent fee the bill proposed for paper bags (while banning plastic ones) as a tax. Supermarket representatives warned the ordinance would make groceries more expensive. But it passed easily, to "lots of cheers," Smith recalls. And "there was a dramatic change" once it came into force. "You didn't see bags flying around, and frankly they weren't landing in my yard, or in my neighborhood."

But three hundred kilometers north in Phoenix, conservative lawmakers weren't happy, and before long they decided to do something about it. The GOP-controlled state legislature approved a law prohibiting local governments from placing restrictions on disposable bags—or containers of any kind. "We didn't want to have this willy-nilly regulation throughout the state," then-Governor Doug Ducey, a Republican, said after signing it. "We think this is more business-friendly." Others had stronger words. One state legislator described bag bans as "the absolute epitome of government run amok," while another slammed them as reflecting "an ideology of collectivism."

By then, David Smith was Bisbee's mayor. He'd left the GOP, and he was furious about the new law. "Because there was no good reason" for the state to stop his town's bag ban, he says, "other than to exert their authority over us." It seemed hypocritical too, coming from members of a party that had long argued for states' rights to be free of unwanted federal interference. "Yet they turned around" and stomped on the prerogatives of "the next level down."

Bisbee believed it had acted lawfully, and refused to rescind its bag ban. A year later, the state upped the ante, passing another law that put teeth—sharp ones—in its earlier measure. It required the withholding of shared tax revenue, money the state typically funnels to localities, from any city or town found to be in violation of Arizona law. "The one thing that I've found seems to get the attention" of local leaders "is a monetary type of notification," the state senate's president said. Even challenging such a finding was risky: Municipalities that did so and lost would have to reimburse the attorney general for legal fees.

With the new law in place, the state threatened to withhold nearly $2 million unless Bisbee repealed its bag ban. "It would have bankrupted us," Smith tells me. Even so, he dragged his feet a little longer. "I was going to make them—the old adage—put up or shut up," he says. "I could not believe that they were going to destroy a city over a plastic bag." But what he thought was a bluff turned out not to be. So the town council, "with advice from our attorneys, caved in," Smith recalls. "We had no choice." To him, it was simple: "They extorted us."

That was 2017. Today, the "desert flags" are again tangled in brush and cactus outside town. "Not as bad as they were, but they're back," Smith says. "If we took pictures before, after, and way after, you could see the bag ban actually worked."

UNLIKE BISBEE, PHOENIX IS BAKINGLY HOT IN EARLY AUTUMN. WITH THE legislature out of session, the state capitol complex—pale stone buildings whose faint reddish tinge echoes the color of the mountains surrounding the city—is all but empty the day I visit. I've allowed plenty of time to navigate security, but the burly guard, in a cowboy hat and white polo shirt, waves me right in. A few minutes later, state Senator John Kavanagh—who sounds just like the Port Authority of New York and New Jersey police detective he was before taking his pension and moving west—sits across from me in a law-book-filled conference room. Catching up with meetings on a recess day, he's dressed casually, in baggy tan pants and beat-up suede shoes, with a thick gray mustache, and he leans back in a big maroon office chair with his hands clasped over his belly.

Kavanagh, a Republican, sponsored the law invalidating Bisbee's bag ban. He rolls his eyes at the suggestion it's hypocritical for state-level conservatives to tie local officials' hands, while themselves resisting such strictures from Washington. "That political argument is based on a total misunderstanding" of government's structure. "The way it is in this country is the states dictate local government powers, procedures, and rules."

To his mind, Bisbee's ordinance encroached on territory that rightly belongs to lawmakers in the capital, and threatened—if other cities followed its example—to create an unwieldy patchwork of local measures. "A lot of these businesses, like supermarkets, they operate all over the state. It's just easier to have the same rules and regulations." And it's not only about bags and other plastic containers, he believes. "If you allow municipalities to go into a lot of areas like this that aren't their purview, you wind up with a whole mountain of laws on all sorts of issues." There are plenty of legitimately local concerns, but bags, in his view, aren't one.

His other big concern about Bisbee's measure, Kavanagh tells me, is that reusable bags are unsanitary. "People just don't clean them," he says, so "they get an awful lot of germs." Together, Kavanagh says, "the health issue" and the need for consistency were "pretty compelling." Plus, he wanted to guard consumers' interests. Plastic bags "are light, they're sturdy," and "if I'm shopping somewhere and I wander into one of these bag-ban zones," the cost of buying a reusable one could be "ridiculous."

When I ask if he has any concerns about plastic bags' environmental impact, he offers up, and bats down, a paper-tiger claim almost no one is making—that Arizona has "a landfill crisis. Well, we don't." And, he adds, even if space had been short, plastic bags weren't the reason. He might have felt differently if Bisbee were near the coast, "but we're pretty far from the ocean, so I don't think any of our plastic Bisbee bags are winding up" on beaches. That, of course, mistakes the real concerns about plastic bags—that producing trillions of them carries a huge toxic footprint, and when littered they can mar landscapes and harm wildlife not just at sea but on land too.

If the state hadn't slapped Bisbee down, it was clear other liberal cities might follow its lead. "The usual suspects would immediately fall into place if we allowed it," Kavanagh recalls thinking. He was right. A few kilometers from the capitol, at a café on the edge of a leafy neighborhood of birdsong and cactus-dotted gardens, Lauren Kuby tells me she began working on a bill to ban plastic bags at retail checkouts, as well as

Styrofoam containers, soon after getting elected to the city council in Tempe, adjacent to Phoenix.

Some questioned the choice, recalls Kuby, a Democrat, who has chin-length blond hair and, when we meet, is wearing cutoff shorts and a red T-shirt, toenails painted blue. Even some allies wondered why she picked plastic bags when Tempe was dealing with big problems like a housing crunch and endless traffic. But with a population of less than two hundred thousand, it was using fifty million of them a year, and spending more than $150,000 annually, she says, cleaning up plastic waste and dealing with recycling delays caused by bags jamming equipment. Plus, a bag ban seemed to her like "gateway legislation" that could show it's possible to solve even intractable-seeming problems. "Get this done," she believes, and people realize going without disposables isn't as hard as they'd imagined.

But Kavanagh's law passed while she was drafting her bill, and it forced her to stop. So when a lawyer asked her to be the plaintiff in a suit challenging the state measure, "I said, 'Yes, I'd be delighted to.'" They didn't get far. A court ruled Kuby wasn't entitled to sue because her bill hadn't been enacted, so the state law hadn't harmed her. The judge acknowledged that left local governments in a double bind—unable to challenge the state edict if they hadn't passed a bag ban, or facing huge financial penalties if they had. "They're damned if they do and damned if they don't," he said.

It was disheartening, Kuby tells me—"such a lost opportunity." The bag ban was "the first of many" political fights she'd wage, but "I took it the most personally, in a way." It was particularly upsetting to talk to lawmakers who were normally allies but feared "poor people were going to be hurt by plastic bag bans," she says. "There's ways to mitigate that," she'd reply—reusable bag giveaways, for example, or free bags for those using food stamps—but she believes lobbyists had frightened her colleagues. Looking back, "I believe I was on the right side of history," she says. "We've only seen the impact of plastic pollution grow."

John Kavanagh feels "totally vindicated" too: His law preserves an option for consumers, he tells me. "And Bisbee still exists. It did not succumb to a mountain of plastic bags smothering all of its residents." One thing they agree on is that local action—if it's not stopped—can be powerful. "Tempe's motto is 'Making waves in the desert,'" Kuby says. Bag bans' opponents saw what was happening there, and in Bisbee, and—just like in Suffolk County, New York, years before—"they were afraid it would spread."

ARIZONA'S "BAN ON BANS" WAS AN EARLY EXAMPLE OF A NEW TACTIC INdustry was embracing—one that would soon gain traction far more widely. Measures that, like John Kavanagh's, bar, or preempt, local governments from acting on a particular issue are known as preemption laws, and they're not unique to plastic. Indeed, while the Bisbee bag fight was underway, Arizona also wrested away Tucson's authority to destroy guns its police seized, requiring they be resold instead. Across the country, states have stopped localities from setting minimum wages; banning fracking; regulating guns, pesticides, or ride-hailing companies like Uber; and guarding LGBTQ rights, among other actions.

Arizona was among the first states to apply that power to bags and other packaging, but others would soon follow. Key to getting such laws on the books was the American Legislative Exchange Council, or ALEC, a low-profile but extremely influential group one headline described as the place "corporations and far-right groups go to buy government policy." Founded in the 1970s by a group of conservative state lawmakers, it was led in its early years by Paul Weyrich, a hard-driving ideologue who became a key strategist of the Christian right's political rise and helped found the powerful Heritage Foundation think tank. Weyrich was frustrated with what he saw as the leftward tilt of national politics, and believed his movement could achieve more in state capitals, which big players often overlooked. By building a network of state lawmakers, Weyrich and ALEC's other leaders realized, they could advance what PBS journalist Bill Moyers later described as their "mission to remake America, changing the country by changing its laws, one state at a time."

Early on, the group focused on social issues, pushing back against abortion rights, gay rights, school desegregation programs, and the Equal Rights Amendment. Eviscerating gun laws was another priority. Its partnership with the National Rifle Association dates to ALEC's 1973 founding, and it began working not long after with the Gun Owners of America, an even more extreme group. The Gun Owners' director, Larry Pratt—ALEC's treasurer in the 1980s—was deeply entangled with far-right militias, white supremacists, and neo-Nazis. ALEC also forged cozy relationships with tobacco lobbyists, even hosting industry representatives at a briefing with President Ronald Reagan.

Despite such achievements, ALEC's leaders knew they needed more money to drive the transformation they dreamed of. So in the 1990s, they restructured. "ALEC must begin to function more like a business," a new blueprint explained. "ALEC's product is policy, and its customers are state legislators and private sector supporters." Companies would pay tens of thousands of dollars for ALEC membership—pocket change for supporters like ExxonMobil, Shell, Coca-Cola, McDonald's, Procter & Gamble, Johnson & Johnson, and Walmart—while lawmakers join even today for just one hundred dollars a year.

Corporate reps get plenty of face time with the politicians at ALEC's conferences, typically held at luxury hotels. Families tag along, and perks include golf, NRA-sponsored shooting, dinners, and cigar parties—the stogies proffered on silver platters—paid for by the likes of tobacco giant

R. J. Reynolds. Lawmakers can apply to ALEC for "scholarships" to offset the cost of attending, or seek funding from home-state lobbyists.

The real business is conducted behind closed doors, at policy "task forces." Each is chaired jointly by a public official and an industry representative, and includes lawmakers and sponsors from companies and conservative organizations. They vote as equals on model bills that—if approved by ALEC's board—lawmakers can introduce word for word in their legislatures. Dues-paying corporations "actually co-authored the bills that ALEC's legislative members took back home and tried to pass," Christopher Leonard wrote in *Kochland*, his study of Koch Industries, the heart of brothers Charles and David Koch's fossil fuel empire—and a key ALEC player. "This partnership appears to be unique in American history, giving companies an unprecedented chance to craft public policy."

Wisconsin Democrat Mark Pocan, who joined ALEC to peek inside, told Moyers's show two main themes defined its agenda: shrinking government and passing "profit-driven legislation" to benefit a particular industry. In his own state, then-Governor Scott Walker's 2011 blitz kneecapping labor unions bore ALEC's fingerprints, as did anti-union bills elsewhere in the country. Another way to protect profits is to limit Americans' ability to sue for harm products cause, an effort that—under the innocuous-sounding heading "civil justice"—is a top ALEC priority. A Missouri state representative cochaired one such task force with a lawyer whose firm represented companies facing hefty asbestos claims. The panel approved bills to limit asbestos liability, and legislators who sponsored them in statehouses called the lawyer as an expert witness more than a dozen times, *USA Today* and the *Arizona Republic* reported. Such laws also hamstring people's ability to sue producers of pesticides, guns, tobacco, chemicals, fossil fuels, opioids, and other drugs.

On criminal justice, for-profit bail and prison companies drive ALEC's work. At one conference, a bail industry representative told lawmakers they were free to tweak his model bill's language, offering to "work with you and your staff on that" and pointing them to a phone number for his group's office, which they could call for help. Obviously his members would profit wherever the bill passed, he acknowledged, but if it helped reduce crime and prison overcrowding—questionable, to say the least—"you don't mind me making a dollar." For-profit bail is illegal almost everywhere other than the United States, where industry lobbying perpetuates a predatory and ineffective system that disproportionately harms people of color. Bail bondsmen still hold a seat on ALEC's board today.

While the copy-and-paste nature of ALEC's model bills may seem shocking, it's not necessarily insidious. Other groups, including the nonpartisan Council of State Governments, share boilerplate legislation too, so states don't have to reinvent ideas. What made ALEC distinctive was the way corporate interests could pay for the opportunity to directly shape government decision-making. As a utility lobbyist outbid by Koch interests in an internal fight over electricity market deregulation put it, author Leonard recounted, "It's a situation where you buy a seat at the table, and then you have the opportunity to vote and drive policy."

Indeed, the Kochs and their vast empire—which produces not just coal, oil, and gas but also petrochemicals such as paraxylene, used to make some of the most common plastics—have long been central figures in the country's right-wing political influence ecosystem, pushing ideas like the privatization of public schools and roads, and the abolition of Medicare, Medicaid, the Environmental Protection Agency, and the Food and Drug Administration. "Like ideological venture capitalists, the Kochs have used ALEC as a way to invest in radical ideas and fertilize them with tons of cash," wrote Lisa Graves, of the Center for Media and Democracy, which obtained and published hundreds of ALEC bills in 2011, drawing a public spotlight to the group's shadowy workings. The family has poured millions into ALEC, and a Koch lobbyist has sat on the group's advisory council for decades.

Driven by the Kochs and other energy and chemical industry supporters, ALEC has put opposition to environmental regulation—and protection of fossil fuel interests—at the heart of its work. Among numerous other efforts, it's offered bills to unravel state renewable power goals, ease pollution limits on coal power plants, promote construction of gas pipelines, and slap surcharges on homeowners' use of rooftop solar panels. Ahead of one meeting, a Koch-backed think tanker promised the chance "to push back against woke financial institutions that are colluding against American energy producers." An ALEC task force soon approved a model bill assailing investors' decisions to eschew fossil fuel holdings as "discrimination," and demanding states keep lists of entities "boycotting" oil, gas, and coal. "ALEC is a big reason the US is so far behind in taking significant action to tackle climate change," a Center for Media and Democracy researcher said.

While ALEC says it does not deny the reality of climate change, its conferences have featured talks—including one titled "Warming Up to Climate Change: The Many Benefits of Increased Atmospheric CO_2,"—casting doubt on established science. Indeed, ALEC's climate views were so extreme, they prompted even ExxonMobil and Shell to leave. Google quit over climate too, with its then–executive chairman Eric Schmidt saying in 2014 that ALEC had been "literally lying" about warming.

The group had already encountered other public troubles. In 2012, the fatal shooting of unarmed Black Florida teenager Trayvon Martin drew a national spotlight to so-called Stand Your Ground gun laws like the one that helped his killer get acquitted. ALEC wasn't behind Florida's law but had helped get similar bills passed elsewhere. NPR reported that the notorious 2010 Arizona "Show Me Your Papers" law, which all but invited police to engage in racial profiling by letting them demand documentation from anyone they suspected of being in the country illegally, was drafted at an ALEC meeting with help from private prison companies that profit from detaining immigrants. And civil rights groups such as Color of Change piled pressure on supporters over

ALEC's backing of voter ID bills. Fearing reputational damage, companies including General Motors, General Electric, Coke, and Pepsi cut ties with ALEC.

Even so, it continues to wield great power—and some of those that withdrew are still represented at ALEC through associations such as the US Chamber of Commerce and a tech industry body whose members include Amazon, Google, Meta, Netflix, X, and Airbnb. Most importantly for ALEC's ability to get bills passed, nearly a quarter of all state lawmakers—overwhelmingly Republican—are members. Its alumni include, at the time of writing, five members of President Trump's cabinet, including the US secretaries of state and homeland security; seven sitting governors; and a fifth of Congress—thirteen senators and ninety-four representatives, including the House Speaker. In one eight-year period, ALEC bills were proposed nearly three thousand times, in all fifty states and Congress. More than six hundred became law. The group, it's clear, has helped create the sharp inequality, overweening corporate power, worker precarity, and public sector decay that define the economic landscape of 2020s America. (ALEC's spokesman did not reply to my emails requesting an interview.)

Given all that, bills grabbing away localities' power to regulate plastic bags sat comfortably in the ALEC matrix, combining antigovernment ideology, protection of powerful industries' profits, and a dash of culture war heat. At a 2014 meeting of an ALEC spinoff for city and county officials, the American Progressive Bag Alliance ran a workshop explaining why bag regulations were "ill-advised and deliberately misleading," wrote one lawmaker who attended. Bag lobbyists were present again the following summer, when—a few months after Arizona passed John Kavanagh's bag law—an ALEC task force approved a resolution legislatures could adopt to express opposition to local laws on bags, boxes, cups, and bottles. "[Insert Jurisdiction]," it instructed, before proclaiming "the free market is the best arbiter of the container." Before long, ALEC's board approved a stronger model bill, borrowing Arizona's language in a measure to expressly forbid municipalities from regulating bags and other containers.

ALEC's vast network quickly kicked into gear, proposing preemption laws in legislatures across the country. Missouri passed one in 2015, and Idaho, Michigan, and Wisconsin soon followed. Minnesota approved preemption in 2017, invalidating a Minneapolis bag ban that was to have taken effect days later. "It was not very good democracy," said the councilman who'd introduced the city measure. The American Chemistry Council plowed $450,000 into Minnesota lobbying in the two years leading up to the law's passage, and the American Progressive Bag Alliance chipped in $60,000. Before long, states including Ohio, Tennessee, Oklahoma, Nebraska, and both Dakotas had also forbidden local governments from restricting bag use.

In Texas, the border town of Laredo had finally passed a bag ban in 2014, six years after lobbyists frightened it into dropping the idea. So Phil Rozenski, an executive at the South Carolina packaging maker Novolex, reached out to a group of local shop owners who'd banded together in opposition. They couldn't afford to do much about the law, but Rozenski, who was also an official at the American Progressive Bag Alliance, said plastic and petrochemical companies wanted to help. "Then Novolex, in association with these other companies, came up with the money," the merchants' group president told the *Houston Chronicle*. The industry had planned to bankroll a lawsuit against Dallas, but the threat alone had prompted the city council there to repeal its five-cent bag fee. Laredo would do nicely instead. "We had a common interest in funding this court case," Rozenski told the paper. (Novolex did not reply to my request to speak with Rozenski. The American Chemistry Council and the Plastics Industry Association, the current name of the old Society of the Plastics Industry, also did not reply to my emails.)

Unlike Dallas, Laredo put up a fight. It had some unexpected bedfellows. Texas cotton ginners supported the ban, because bags tangled up in the crop were damaging machinery and reducing their product's value. But plastic-industry dollars enabled the merchants' group to hire a new lawyer, and in 2018 the state supreme court said the ban violated a 1993 Texas waste law. The ruling, which took down eleven other local measures, made preemption the law in yet another state. Today, local bag measures are prohibited in seventeen states. As one headline put it, "Plastic Bags Have Lobbyists. They're Winning."

It's not hard to see why industry embraced preemption. While an individual bag has little value and less weight, the mountains produced every year add up to a lot of plastic—and a great deal of money. Globally, the market for the thin, classic "T-shirt bags" was worth more than $7 billion in 2023, and it's expected to climb past $9 billion by the early 2030s (and T-shirt bags are just one segment of a larger plastic bag market, including resealable zipper bags, garbage sacks, and heavier shopping bags, that is expected to grow from $24 billion to $34 billion in the same period). What's more, both advocates for and opponents of plastic bag restrictions agree they're just a first step. If bag usage can be sharply reduced without noticeably harming quality of life, consumers might start wondering whether the same might be true of packaging, disposable utensils, and other items they'd thought of as impossible to part with. Stephen Joseph, the California "patron saint of plastic bags," had understood that, warning in 2013 that his opponents were attempting "the stigmatization of all forms of plastic. The entire plastics industry needs to wake up, work as a unit, and fight like hell."

BACK IN NEW YORK, STEVEN ENGLEBRIGHT KEPT WORKING ON PLASTIC after becoming a state assemblyman in 1992. He joined a colleague's fight to expand the state bottle bill, whose deposit requirement originally applied only to carbonated drinks. In the end, they were able to add only water bottles—a significant win, but disappointing nonetheless. It was "the same pattern" he'd seen at home in Suffolk County, he tells me, "industry basically doing everything they can to frustrate the creation of laws" restricting throwaway packaging.

Still, he kept pushing. Progress got easier after Democrats took control of the state senate. By then, Englebright was chairman of the assembly's environment committee. In 2019, more than thirty years after his bag ban passed in Suffolk County, New York followed California to become the second state in the country to enact a statewide plastic bag ban. To Englebright, the achievement feels bittersweet. He can't help thinking about what might have been in the decades between his initial success and the enactment of the state's more limited law. "It doesn't sit well," he tells me. "In those intervening thirty years, the Great Pacific Garbage Patch was created," and microplastics spread not only to every corner of the planet but throughout our bodies too. "Could we have avoided all of that?" he sometimes wonders, if his law had stood and been copied elsewhere. "I think we could have avoided the prevalence" and ubiquity. Companies said they wanted to be part of the solution, but Englebright can see "they had no intention" of reducing production. "Instead, it's been the other direction."

Today, he's still fighting the battle he began all those years ago. He lost his state assembly seat in 2022, when he was seventy-six. Instead of retiring, he ran for his old job in the Suffolk County legislature, and won. "I'm still on top of the turf, not under it," he tells me with a laugh. "I still think I have something to contribute." When we speak, he's working on a bill to require that take-out food sellers provide plastic utensils only if a customer requests them. With Republicans holding a supermajority in the legislature, he doesn't know if it will pass. But, he says, "I can tell you this: If the bill fails, I'm going to reintroduce."

Others are equally determined. In the face of relentless industry opposition, twelve states and hundreds of American cities and towns now bar retailers from providing free plastic shopping bags, and some of their laws have tackled other single-use plastics too. In 2021, Maine became the first state to require companies to contribute toward the cost of recycling programs, and Oregon soon followed. California passed an even more ambitious law in 2022, mandating a 25 percent reduction in plastic packaging, requiring all single-use packaging be recyclable or compostable, and forcing manufacturers to not only pay for recycling but also cough up $5 billion for a fund to mitigate plastic pollution's damage to human health and the environment.

But in addition to showing what's possible, the years of fighting that won passage of such laws have also demonstrated a more painful truth. Everywhere, industry has wielded the same weapons—campaign cash and endless lawsuits, for starters. When those haven't succeeded in stopping restrictions, lobbyists have pushed to riddle them with loopholes that have made new laws less effective, if not counterproductive. They've created anodyne-sounding groups like Californians for Recycling and the Environment—founded by two executives from packaging maker Novolex—to warn of "a future without toothpaste, baby formula and dog food" if the plastics law passed. Often, they've funded their own studies, and selectively, even misleadingly, quoted other research to confuse consumers about plastic's harms.

By putting so much money and muscle into that fight, industry has succeeded at something bigger than just protecting bags. It's prevented—or at least slowed—action on other single-use plastics, and the safety of the chemicals used to make them. What could have been a steady march toward a more sustainable world "was diverted, stalled, and [its] momentum broken by what this industry behavior has done," Englebright laments. "To make a long journey, one must make many, many steps. And we've had too few steps."

CHAPTER 9

"No Silver Bullet"

2009—Boeing's 787 Dreamliner enters service. Plastic composites make up 50 percent of the plane and all of its outer skin. One headline calls it "Boeing's Plastic Dream Machine."
315 million metric tons of plastic produced.

With far less drama than across the Atlantic, the European Union moved in 2015 to sharply cut plastic bag use. Each member country, the EU decided, would have to require fees be charged for bags or find another way to reduce per-capita consumption by 80 percent. The idea enjoyed broad public support, and industry opposition was far less intense than in the United States. "We pushed through an open door, in a sense," one activist told me. As the new rules came into force across the continent, the bags' presence steadily fell, with billions fewer used each year.

With that done, Europe's leaders were ready to go further. Officials drafted a new proposal, aimed at tackling marine plastic pollution, and published it in 2018. The timing was fortuitous. Disturbing images of that pollution's damage—a biologist extracting a straw from a sea turtle's nostril, albatrosses whose bodies had decayed into beaks and feathers heaped around pieces of brightly colored plastic they'd swallowed—were racking up millions of views online. It sent anxiety about plastic soaring and created political momentum for the new effort, which focused on the ten categories of single-use items most often ending up in oceans. Each would be subject to a restriction of some sort. Throwaway plastic utensils, plates, stirrers, straws, and cotton swabs were banned outright; they could be replaced with reusable versions or made from other materials, like cardboard. For cigarette butts, the proposal required tobacco companies to cover collection and disposal costs.

The plan wouldn't reduce plastic bottles' ubiquity, but it had the potential to transform both their production and their end-of-life fate. It mandated that by the end of the 2020s, 90 percent of used plastic bottles be collected; new ones would have to be made from at least 30 percent recycled material. Such requirements are necessary to ensure PET—the most recyclable type of plastic—actually gets recycled. Without them, recycled plastic can't compete with virgin material.

Disposable cups were trickier to manage. The EU plan—it's known as the Single-Use Plastics Directive—covered even those that, like most take-out coffee cups, are made from paper or cardboard coated with an invisible layer of polyethylene. Strictly regulating them seemed like a stretch, so the proposal took a softer approach, requiring each European nation to achieve "an ambitious and sustained reduction" in their use. Along with other items whose plastic content might not be obvious—period products and wet wipes—the cups would have to carry a label disclosing the material's presence. The image eventually chosen showed a dead turtle floating amid plastic trash.

The plan sailed through the European Parliament, a 705-member body of lawmakers elected by voters across the continent. Next it would have to win approval from the Council of the EU, made up of representatives from the governments of all twenty-seven member nations. For companies loath to openly oppose "this thing that pretty much every person on the street thought was a good idea," that relatively private setting looked like a more promising place to win concessions, one expert told me. So while publicly proclaiming support for reducing plastic pollution, industry pushed behind closed doors to oppose, weaken, or delay many of the directive's provisions. Among other things, lobbyists pressed to get items off the restricted list and to ease the bottle collection targets and recycled content mandates, said Corporate Europe Observatory, a group that tracks lobbying in Brussels, the EU capital. Industry was particularly vociferous in opposing a requirement that caps be tethered to bottles to reduce their likelihood of becoming litter.

A tranche of documents obtained by an *Irish Examiner* journalist and a former European lawmaker offered a glimpse of the channels through which lobbyists could make themselves heard. As Ireland's delegation in Brussels reviewed the draft directive, officials back in Dublin were in close contact with groups representing the food, beverage, retail, and packaging industries. In one email to an alliance whose members include Coca-Cola, Unilever, Kellogg's, and the UK-based supermarket chain Tesco, an Irish official pasted part of an EU draft and suggested a "call within the next 30 minutes" to discuss it. In another message, an official wrote they wanted to "run the following text [changes] by you," and shared language on the methodology for calculating compliance with the bottle collection target, a key industry concern. "Feel free to give me a call if you want to discuss," an official wrote in an exchange about the directive's definition of single-use plastics, promising to "keep you posted on any developments."

While Dublin ultimately backed the Single-Use Plastics Directive and called for even stronger measures, its council submissions also conveyed companies' views: "We would like to point out that Industry in Ireland has indicated concerns" about various provisions, one memo said, going on to detail them. Other European governments were almost certainly as cozy—if not more so—with industry, Corporate Europe Observatory, the lobbyist tracking group, said in analyzing the emails.

In the end, industry didn't get much for its efforts. The Council of the EU approved the directive, and the final version gave away little more than deadline delays. All told, the Single-Use Plastics Directive went from proposal to settled deal in seven months, far quicker than the two years typical for a bloc whose sometimes unwieldy legislative process reflects the need to balance the priorities of countries whose politics often pull in competing directions. Tatiana Luján, an environmental lawyer I met, made a swooshing sound—and a swooping motion with her hand—as she recalled the measure's glide path to enactment. The relative lack of contentiousness was visible, too, in the demeanor of the officials and lawmakers involved. "They were so hopeful," she said. "They were happy. They knew people were asking for this and it was a good thing they could achieve."

It wasn't over yet. Officials still had to draft the detailed regulations that would govern rollout of the directive, and industry quickly shifted its focus to pressing for advantage there. Technical questions like exactly how to calculate recycled content seem abstruse, but it's in such nitty-gritty that lobbyists can shape what a law means out in the world, and what companies must do to comply. On every detail, recalled Luján, they fought vehemently to water down requirements. A battle over how much plastic content meant an item was covered by the directive "dragged on for two years," she said, with industry arguing that "if it has less than 10 percent" it shouldn't be included. That wasn't what the directive had said; because even a small amount of plastic makes a paper or cardboard item unrecyclable, the dead-turtle label applied no matter how low the percentage.

The new rules are proving powerful. To achieve the 90 percent plastic bottle collection target, many of the European nations that didn't already mandate deposits are now doing so. And some countries are going even further than required. Despite the bumps along the way, the directive's relatively smooth path to enactment reflected a European relationship with environmentalism—and the use of regulatory power more broadly—that is dramatically different from America's. As an American living in Britain, I've long thought of the EU as a far savvier and more effective regulator than either of my two countries—a place where political leaders are able to prioritize public good over private profit, even when it means angering vast corporations. And there's plenty of truth to that. Even so, I'd learn on a visit to Brussels that industry gets its way there—if not quite as frequently as in Washington or London—much more often than I'd imagined from afar.

THE EUROPEAN PARLIAMENT'S VAST BRUSSELS OFFICE COMPLEX IS sleek, modern, and bright. Continent-wide elections are looming when I visit, and big blue signs out front urge passersby to "Use Your Vote." Inside, at the top of an escalator leading up from street level, I see a moderator chatting with guests in an open-sided TV studio, and peer down spacious, window-lined walkways that lead off in several directions. The place is bustling, with gaggles of visitors taking tours or advocating for causes, and a multitude of languages float behind them. This is where, in late 2022, a proposal to regulate disposable packaging by the European Commission, the EU's executive branch, landed for lawmakers' consideration. The plan, known as the Packaging and Packaging Waste Regulation, or PPWR, was much farther-reaching than the Single-Use Plastics Directive, and it held the potential to transform the way many companies did business. Industry's previous lobbying pushes would pale in comparison to the one it unleashed this time around.

After buying coffee from a barista in formalwear, I grab a seat at a busy café beside the escalator and look around for Grace O'Sullivan, a lawmaker from Ireland. A woman in a green blazer is sitting at a table nearby, and—although I don't know whether the color nods to her Irishness or her Green Party affiliation (or neither)—my hunch that she is the legislator I'm scheduled to meet is correct. O'Sullivan is chatty and engaged, but we're just a few minutes in when her assistant, sitting beside us, looks up from her phone. "I'm sorry to interrupt," she says. "Our prime minister has just stepped down." As O'Sullivan hurries away to deal with the upheaval back home, the young aide, Rose Ní Chléirigh, stays to talk. O'Sullivan was her party's lead negotiator on the packaging bill, and as her point person on the issue, Ní Chléirigh spent more than a year deep in its weeds, haggling over changes with other staffers in hours-long meetings.

Ní Chléirigh is stylish in cropped black trousers and a gray ribbed sweater, with a broad smile and straight brown hair. She talks at high speed, but in full, thoughtful paragraphs, explaining the complex bill's ins and outs. The original plan, in her view, was powerful enough to bring "far-reaching, systemic change" to packaging. If passed with its most ambitious provisions intact, it would begin to foster new (or, really, old) systems built around reuse rather than disposability. "The proposal was really brave," she told me. "It could have really fundamentally tackled throwaway culture, and this convenience at any cost that we've become so used to." Those, anyway, "were the hopes and dreams."

In introducing its proposal, the commission noted every European, on average, generated nearly 180 kilograms of packaging waste a year, a number that had been rising steeply, and would, without action, climb by another 19 percent by decade's end. Its plan aimed not just to slow that growth but to reverse it, cutting packaging use by 5 percent by 2030 and 15 percent by 2040. Doing so would shrink climate-warming emissions, save water, and avert more than €6 billion (about $6.8 billion) in environmental damage, the commission said.

Food service was a key focus. The draft proposed to bar eateries from providing single-use dishware, utensils, and condiment packets for customers dining in. A few years ago, that might have seemed unremarkable, but recently—in Europe, as elsewhere—the practice, once confined mostly to fast food, has become widespread. The original PPWR also proposed binding targets that would push the sector away from throwaway packaging for takeout. Restaurants and cafés would have to build or join systems for getting back and reusing food containers and cups. By 2030, 20 percent of to-go cups and 10 percent of take-out food boxes would have to be reused, with the targets rising in 2040 to 80 percent of cups and 25 percent of food containers. While companies could choose how to reach them, supporters believe those quotas would likely spur the development of pooled systems in which restaurants share reusable boxes, and customers who choose them when ordering hand them back when getting their next delivery or visiting a participating business.

It's not as far-fetched as it might sound. Start-ups providing reusable cups and food boxes, along with facilities to wash them, are already rolling out such efforts in Europe and beyond. In Denmark's second-largest city, Åarhus, for example, more than forty cafés and bars are sharing forty thousand cups, in two designs, for hot and cold drinks. Customers return them—and get their deposit back—at reverse vending machines, and the company running the system washes the cups and redistributes them to retailers. It hopes to soon expand into take-out food containers.

Setting a legal target, supporters believe, would make such systems common. It's not about requiring individuals "to be wandering around with a whole box of containers," one advocate explained, but creating infrastructure that makes reuse accessible and easy.

Among its other provisions, the draft bill also proposed to ban some types of single-use plastic packaging outright—the tiny toiletry bottles provided in hotels, for example, and throwaway wrapping for fruits and vegetables (except where necessary to protect the produce). A quarter of packaged drinks would have to come in reusable containers by 2040 (wine got a lower quota). Online retailers like Amazon would be required to make half their packaging reusable. Ninety percent of packaging for big appliances like refrigerators would have to be reused. And by 2030, all packaging would need to be recyclable, with at least 70 percent actually getting recycled. The recycled-content minimums previously created for plastic bottles would be introduced for all plastic packaging, and new rules would restrict use of heavy metals and other dangerous chemicals. Anticipating industry's economic arguments, the commission said its plan would create more than six hundred thousand jobs and save Europeans one hundred euros a year each if companies passed savings on to customers.

The pushback was ferocious. As Ní Chléirigh negotiated changes to the bill, she was bombarded by "more emails than I can fathom," she tells me. "A real onslaught." If she didn't reply, senders followed up by phone. "I started this with a nice spreadsheet of all the people who had contacted us," she recalls, but "in the end, I had to just accept I could not keep on top of" it, and other lawmakers' aides reported similar experiences. All the sectors that would be affected by the rules—restaurants, retailers, food and drink companies, the manufacturers of packaging, and the makers of materials that go into it—had opinions, and often their priorities conflicted. "Everyone said, 'Oh my God, the lobbying on this has just been more than I've ever seen,'" Ní Chléirigh recalls. "There were so many companies, so many different demands."

Lobbyists lurked everywhere, says Gerrit Krause, a rumpled aide with pushed-up sleeves and a reddish beard who works for lawmaker Delara Burkhardt, a German Social Democrat. "I got approached in the canteen and the cafeterias of Parliament," and cafés nearby, by industry representatives keen to discuss the bill, Krause tells me. At bus and metro stops near the legislative complex, posters blared companies' talking points. Industry placed paid articles trumpeting its views in *Politico*, whose Brussels coverage is widely read by EU decision-makers. An alliance of fast-food chains including McDonald's, Dunkin', and Yum!—the parent company of KFC, Taco Bell, and Pizza Hut—adopted the slogan "Together for sustainable packaging." To Pascal Canfin, chairman of the

Parliament's Environment Committee, that was positively Orwellian—"the exact opposite of what they were doing."

One night, lobbyists from a group whose members included Starbucks and the packaging makers Amhil and Novolex put door hangers warning the bill would "End Takeaway Food & Drink" on the entrance to every Parliament member's office. Beneath a photo of a young mother eating from a clear plastic container, the glossy flyer urged lawmakers to "Save Our Takeaway!" by backing amendments that would riddle the bill with loopholes. The suggestion takeout might become impossible frightened many politicians, and correcting those misconceptions took a lot of work, Krause says. He and his colleagues explained "why it's not as dangerous as these companies are suggesting," emphasizing, for example, the plan's exemptions and long transition periods. "Myth busting," he tells me, was "quite time consuming."

One of the biggest players in the pressure campaign was McDonald's. The requirement to provide reusable tableware for dine-in patrons would have forced it to change its long-standing model of serving most everything in throwaway boxes, wrappers, and cups. The company commissioned a study titled *No Silver Bullet*, which concluded reuse was not necessarily the answer, even for on-site dining, in part because it would replace throwaway paper packaging with thick, durable plastic and introduce the need for dishwashing. "This well-meaning regulation," the burger giant said, would "be worse for the environment," increasing greenhouse gas emissions and water consumption. "The idea of reusing something over and over again seems the obvious solution—but it's not so simple," Jon Banner, McDonald's sustainability chief, wrote in an advertorial the company bought in *Politico*. France had already introduced a similar requirement on its own, and Banner explained that customers there often kept or threw away plates and bowls intended to be left at the restaurant. Few people returned cups to Dutch and German outlets experimenting with deposits for reusables. "A rush to a solution for a complicated situation will only make the problem worse," he argued. Recycling was the better answer, Banner contended, and McDonald's was working on technology for processing items contaminated by food and those—like plastic-lined paper cups—made from more than one material. Reading that, I couldn't help thinking of Styrofoam makers' long-abandoned promises to recycle that material when Steven Englebright was trying to ban it in Suffolk County, New York, back in the 1980s.

Another industry-commissioned report offered findings similar to McDonald's. But "the problem with these studies is they don't withstand scientific scrutiny," Krause told me. Indeed, more than fifty experts, in an open letter, warned they contained methodological problems that could have significantly skewed their conclusions. Both studies were "at best flawed, at worst deliberately misleading," one of the signatories wrote. Canfin, the Environment Committee chairman, a French centrist, said the McDonald's report helped create the false impression that business decisions based on self-interest were actually grounded in science—"exactly the same method that the tobacco industry used" decades ago.

And as it did for so long with cigarettes, that strategy worked, Canfin told me, creating doubt, in this case, about an official assessment that found even thick plastic reusable items were less harmful than throwaway paper packaging. Lawmakers not immersed in the details got "the impression that actually it's not a fact, it's an opinion," he said. Baseless studies "managed to change the narrative."

McDonald's had powerful allies to help it capitalize on the confusion it sowed. Massimiliano Salini, a lawmaker from Italy representing right-leaning legislators in negotiations on the packaging proposal, hosted a launch event at Parliament for the *No Silver Bullet* report. "The best sustainability solutions come from industry, not governments," he said, adding that he was working on changing the bill "because we consider it profoundly wrong to privilege the reuse model at the expense of the recycling one."

Salini wasn't the only Italian who saw it that way. Packaging makers have a major presence in the country, and they pressed Italy's representatives to do their bidding. The Naples-based conglomerate Seda is one of McDonald's major European suppliers (along with Huhtamaki, based in Finland, another country whose representatives pushed hard against the most stringent proposals), and its president also runs the European Paper Packaging Alliance, which played a big role in the lobbying effort. Italy, declared Prime Minister Giorgia Meloni, "does not give in to solutions that penalize our industry."

The Italians came through. An aide at the European Parliament shared emails with me in which industry groups—manufacturers, paper producers, winemakers, and others—suggested detailed changes to the legislation. Italian lawmakers from across the political spectrum offered those changes, word for word, as amendments to the bill. They proposed axing the ban on single-use packaging for eat-in meals, deleting the reuse targets for takeout, letting recycling get companies out of similar mandates in other sectors, and eliminating the bans on specific types of packaging. That last change had been suggested in a position paper signed by Plastics Europe—the main European plastic producers' association—and an alliance of packaging makers and users whose members included ExxonMobil Chemical, Chevron Phillips Chemical, INEOS, Dow, Amazon, PepsiCo, Nestlé, L'Oréal, and Unilever. (Plastics Europe declined my request for an interview, as did Salini's office. Five other industry groups representing plastic and packaging makers and food and drink companies also refused or did not respond to my queries.)

Maria Angela Danzí was one Italian who wasn't playing. A lawmaker from the country's Five Star Movement, a populist left-right hybrid, she won her seat in the European Parliament just before the packaging proposal arrived. When I meet her at the Parliament's café, she's in black jeans and high heels, with neatly coiffed blond hair and a Gucci handbag. We head to her office, where an assistant translates as Danzí tells me of her shock at seeing, in a bar reserved for lawmakers and their guests, a pair of lobbyists displaying an array of take-out cups—like old-fashioned traveling salesmen, she says—to politicians. "I was astonished and disappointed." I ask about another lawmaker's allegation that lobbyists had followed his colleagues into bathrooms. Danzí didn't see that, but "they were like flies," she tells me. "They were everywhere." Other Italian lawmakers warned Danzí against picking fights with such powerful interests. But she was so rattled by the heavy-handed tactics, she asked the Parliament's president to investigate. "Now," she tells me, "some colleagues hardly say 'ciao'" when she sees them.

THE BILL WAS EVISCERATED. FROM THE START, THE ENVIRONMENT COMmittee chairman Canfin recalled, maintaining the initial level of ambition looked difficult. First to go were the reuse targets for takeout. Beyond the Green Party, there was little support for them, "because it's obviously very complicated to organize," he told me. "The meat came off" in other areas too, as one activist put it, describing the bill that emerged from Parliament as "outrageously bad."

But the next step in the lengthy legislative process restored some of what was lost. At the Council of the EU, the jockeying among national governments' representatives typically weakens bills, but in the case of the packaging measure, that body put some of the meat back on. The finished product was not nearly as powerful as the original had been, but it was far stronger than what the heavily lobbied Parliament had approved. There were some clear wins, including new restrictions on the use of dangerous PFAS forever chemicals in food packaging. And the law retained its headline goals for reducing packaging waste, although without a plan for how to achieve them. On take-out containers, it required vendors only to offer a reuse option, saying they should "endeavor" to serve 10 percent of to-go food and drinks in reusable packaging by 2030.

The vision of a future in which reuse replaces disposability was mostly gone, "dramatically watered down," Canfin said, in favor of industry's preferred solution of recycling. "The way agreement was found was by generating huge, massive loopholes and exemptions," one advocate told me.

But the law did take direct aim at plastic. Some of the strict measures once intended to apply across the board now covered just this one material. It was an easier lift politically, the German legislative aide Gerrit Krause explained, because Europeans have come to see plastic "as the bad guy," while paper and cardboard have "this reputation of being kind of natural." Indeed, Canfin told me, once negotiators said "OK, fine. We give in" on paper, the bill was suddenly "very much easier to sell."

So restaurants could keep using throwaway packaging for dine-in customers, as long as it wasn't plastic. A similar spirit was brought to packaging used to ship goods to consumers or between businesses. While cardboard boxes were unrestricted, plastic would, for the most part, have to achieve reuse targets. Single-use plastic wrappings were barred for produce, as were the miniature hotel toiletry bottles.

Plastic brings some unique harms, so reducing its use is a real step forward. But simply replacing it with something else fails to address the rapacious harm of disposability itself. Producing ever-greater amounts of paper and cardboard drives destruction of forests and carbon-rich peatlands from Brazil and Indonesia to Sweden and Finland, releasing climate-warming greenhouse gases and threatening wildlife that depend on those habitats. And because paper-based packaging—particularly for food and beverages—typically contains some plastic and chemical additives, it's hard to recycle, so it often ends up landfilled or burned.

The original plan had reached for something big, aiming to change a model that, in recent years, has become baked into everyday life, and much of the economy. In the end, the law was half a loaf. "There's some really good stuff in there," Rose Ní Chléirigh, the young Green Party aide, tells me. "But on the whole, it's failed to be the transformative instrument it could have been." So while she was glad industry "didn't win on all fronts," to her mind what Europe delivered "doesn't reflect the urgency of the environmental crisis we're in."

That duality—a law that marks a major step forward while still not doing what the moment demands—reflects the reality that disposability is too profitable to too many powerful interests for us to easily wrench ourselves free of it. I wasn't wrong to think, before my visit, that Europe has shown more seriousness about confronting planetary peril than just about any other major political entity. But the painful truth is that that's a low bar. The EU had gone a lot further on reining in plastic than the United States, but even so, the muscle of industries invested in the status quo was blocking urgently needed change.

The packaging proposal had also run headfirst into a changing political reality. Across Europe, the late 2010s had been a moment of ascendence for environmental action, as young climate activists pushed politicians to get serious about the crisis. But by the mid-2020s, with anger rising over the soaring cost of living, the right and far right were surging, their representatives simultaneously fanning and riding a cultural backlash against green measures some disaffected Europeans saw as the unaffordable preoccupations of an out-of-touch elite.

That wave slammed into Brussels not long after my visit, when 2024 elections dramatically expanded right-wing populist parties' presence at the European Parliament. The Green Party took a particularly big hit. Among those losing their seats was Grace O'Sullivan, whose assistant, Rose Ní Chléirigh, had worked so hard on the packaging bill.

Even before her boss lost her job, Ní Chléirigh told me it had been dispiriting to watch corporate money get its way despite many voters' desire for action on a tangible problem they could see up close. Early on, she recalled, the bill had seemed like "such a no-brainer," because people "really relate to 'Oh yes, less packaging—good thing.'" It turns out, she said, "that doesn't matter" in the face of big industries' resistance. "I know I'm naive, and this is my first job in a political environment like this," Ní Chléirigh told me. "But I really was shocked at just how it really is a case of 'If you are big, powerful, and rich, you get your way.'" To Tatiana Luján, the environmental lawyer who described EU leaders' joyous expressions as they enacted their earlier plastics rules, the stiff resistance this time around was no surprise. Disposability, she reminds me, allows businesses to push the expense of dealing with packaging waste onto the public. If they had to create systems for reusing it instead, "that would put the cost on them." So the throwaway model amounts, in her view, to "a free ride" that pads profits. "Of course they want to protect that."

CHAPTER 10

"An Impatient Billionaire"

*2011—One million plastic bags are used every minute.
343 million metric tons of plastic produced.*

Jim Ratcliffe developed his formula for business success decades ago, and he applies it with gusto. It's made him one of Britain's wealthiest men, and turned his company, INEOS, into a leading global petrochemical concern. The playbook goes like this: Find chemical plants and refineries competitors are running badly and eager to unload. Buy these unloved assets cheap, financing the deal with lots of debt. Cut costs fast: Jobs and corporate credit cards are good places to start. Lots of plants operate below capacity, and revving them up to maximum output boosts profit. Earnings should double in the first five years. Under Ratcliffe's hard-driving management, they usually do.

Ratcliffe—Sir Jim, since he was knighted in 2018—is tall and lanky, with a long, craggy face and hair a tad on the shaggy side. He prides himself on informality, and loves extreme sports and over-the-top adventure. He's trekked to the North and South poles, completed the Ironman triathlon and a 240-kilometer ultramarathon through the Sahara desert, driven from Mongolia to Beijing, and celebrated his sixtieth birthday with a 10,000 kilometer motorbike ride across six African countries. He broke his foot along the way, but stuck a ski boot on it and kept going. "I can't think of anything I set out to do and have not completed," he's said.

The son of a carpenter and a secretary, Ratcliffe grew up in an old mill town just outside Manchester, living in public housing until he was ten. Even then, his dreams were big. "I used to think I would rather like to become a millionaire," he once recalled. Winning a spot at one of northern England's best state schools was an early step toward that goal, and he soon became the first in his family to attend university, earning a degree in chemical engineering. He got a job at a BP chemical plant but was fired after three days, when the company learned he'd had eczema, a sign he might be unable to tolerate the chemicals on-site. Years later, he'd joke that he only got to work at a BP plant after he bought one.

His first acquisition from the oil giant came in 1992. Ratcliffe was working in venture capital by then, and dreaming of running his own company. The story of INEOS's origins and subsequent growth has taken on the patina of business legend. He was just thirty-nine when he and a partner, John Hollowood, put together financing to buy one of BP's chemical businesses. Hollowood was wealthy, but Ratcliffe remortgaged his house to come up with £140,000, his piece of the approximately £37 million purchase price. "I put all my chips on it," he later recalled. "If that had gone down, I'd have been in a mess." It didn't come to that. Within two years, the business had more than doubled in value. "Jim was eager for success," Hollowood later told the *Financial Times*, "very, very stubborn, and very determined."

The pair called their new company Inspec, but it was just Ratcliffe's first step—there were more big gambles ahead. One of the most consequential came a few years later. When Asian economies nosedived in 1997, advisers urged Inspec to sell a plant it owned in Antwerp, Belgium. The facility produced chemical raw materials most investors found unappealing, but Ratcliffe saw potential at a bargain, and he decided to split from Inspec and buy the complex for himself. It turned out, one of his financiers observed later, to be "a really smart deal."

The Antwerp plant would become the foundation for his new petrochemical empire. He called it INEOS. As it grew, Ratcliffe did what he does best—streamlining inefficient operations, clearing bottlenecks, and wringing every penny possible out of a plant. When INEOS takes charge of a new facility, it cracks down hard on spending. "It's a short, sharp shock tactic," longtime colleague Tom Crotty explained. "Some people get it, some don't, and the majority are confused for a short time before adjusting."

Ratcliffe put his formula to work again and again as he expanded his company's holdings, borrowing against assets he already owned to grab new ones. "We mopped up unfashionable, unsexy commodity and bulk chemical businesses" at low prices, he wrote in a memoir of INEOS's history. "His negotiating tactics are not loved by everyone and people don't like it when pain is inflicted," one contemporary said, but such relentlessness tended to pay off handsomely.

INEOS was building toward its biggest purchase yet, one that would prove transformational. Innovene, a BP subsidiary, was the world's fifth-largest petrochemical company, with nineteen manufacturing sites across the US and Europe. Its annual revenues were $22 billion—more than three times INEOS's. BP wanted to offload it, and was preparing to float Innovene on the stock market. Ratcliffe had a different idea. Uncomfortable with the loss of control issuing stock entails, he has always owned INEOS privately, together with two colleagues. While Innovene's top executives were readying their public offering, a small circle of BP leaders were secretly negotiating a direct sale to Ratcliffe. He took a critical call while cycling in the Scottish Highlands, deciding on the spot to up his offer by $1 billion, not mentioning "I'd just got off my mountain bike, covered in mud and shit," he recalled. The purchase quadrupled INEOS's size overnight. The guppy, one industry expert commented, had swallowed the whale. INEOS saw it differently: "We were a lean, mean, fast-moving barracuda swallowing a whale," one executive said.

At the last minute, BP had added two big oil refineries into the package, and as Ratcliffe later said, "We knew fuck-all about refineries." He was unfazed. A friend sent him a link to HowStuffWorks.com, and the joke that INEOS's leaders learned what they needed to know about refineries from the site entered company lore. It was typical of a corporate culture one adviser—likening Ratcliffe's vision to legendary investor Warren Buffett's—described as "incredibly unstuffy and unbureaucratic." Some prefer the word *ruthless*. "You want to do business with him, but you also feel a little bit nervous," said a banker who worked with INEOS, "because of how competitive and aggressive they are."

Ratcliffe certainly isn't afraid to play rough. With sales tanking after the 2008 financial crisis, his debt-laden empire was in trouble. He pressed Britain's Treasury to defer a £350 million tax payment, and when then–Prime Minister Gordon Brown refused, he moved INEOS headquarters to Switzerland. (The company returned to Britain in 2016, when the Conservatives were in power, cutting corporate tax rates.) Later, he argued passionately for Britain's exit from the European Union but seemed to see no contradiction in moving his own tax residence to Monaco. "I paid my taxes for sixty-five years in the UK," he said. "And then when I got to retirement age, I went down to enjoy a bit of sun. I don't have a problem with that, I'm afraid." He is not retired.

One of the oil refineries INEOS had inadvertently picked up in the Innovene deal was in Grangemouth, Scotland, part of a site that also included a major petrochemical operation. Like other British and European chemical facilities, Grangemouth had a serious problem. While America's fracking boom had unleashed a torrent of cheap ethane there, Britain's once-rich reserves in the North Sea were dwindling. Without sufficient volumes of its main raw material, Grangemouth's cracking complex was running at just 40 percent of capacity. Ratcliffe had seen up close how abundant ethane had transformed the US petrochemical landscape—cheap fracked supplies more than tripled profits at an INEOS plant near Houston.

He hoped to replicate the experience across the Atlantic by fracking in Britain. But his determined campaign to do so failed, stymied by concerns about toxic pollution, earthquakes, and climate change. So he came up with another idea: He'd bring American ethane to Europe. The technical challenges of transporting huge volumes of it such a long distance were daunting—"unprecedented, and almost inconceivable in the scale of its complexity," the INEOS memoir declared. "Our [industry] colleagues in the States didn't think for a second that we would ever be able to pull this off," said the manager Ratcliffe put in charge of the project.

Before committing to spend the £1.5 billion (about $2 billion) necessary to realize his vision of a "virtual pipeline" of ships carrying American ethane to Europe, Ratcliffe was determined to eliminate a bothersome stumbling block. In 2013, the union representing Grangemouth's workers was resisting his proposal to freeze wages and scale back the pension plan, a bitter standoff that was Britain's biggest labor dispute in decades. As the battle escalated, the union vilified him as a tax exile on a yacht in the Mediterranean, and headlines called him Dr. No, after the Bond movie villain. In his view, he was simply doing what was necessary to turn around a plant that was hemorrhaging money. The idea "that you are a bad guy because you have a loss-making plant and want to shut it down is flawed thinking," he said.

Ratcliffe had a powerful card to play. If the union didn't accede to his conditions, he announced, he would not fork out the £400 million needed to prepare Grangemouth to receive imported ethane. Instead, he'd shutter the plant for good, at a cost of eight hundred jobs—a blow the area's parliamentary representative said would "destroy a large part of my community." Workers quickly gave in, a painful defeat.

With his union problems solved, Ratcliffe secured contracts for not just ethane but also propane and butane from Range Resources, a big fracking firm drilling in southwestern Pennsylvania. He hired a pipeline company to upgrade and expand a line that would carry the supplies to a port near Philadelphia. And he had that facility completely overhauled, with huge tanks to chill the gases to minus ninety degrees Celsius so they could be shipped as liquid. "The fact that ships large enough and specialized enough" to haul the stuff didn't exist "simply meant that INEOS would have to build them," the company's official history boasted. It ordered eight custom-made supertankers—each the length of two football fields—from shipyards near Shanghai and dubbed them the "Dragon Ships."

The first one landed at INEOS's Norwegian plant in March 2016. A few months later, a bagpiper played onboard as Grangemouth's inaugural load approached a dock where Ratcliffe and three hundred guests were assembled. After all the preparatory fanfare, high winds forced a last-minute postponement; the ship couldn't dock until they subsided. But the wait would be worth it. American ethane enabled INEOS to crank the Grangemouth complex back up to full capacity. Now, it produces a million and a half tons of petrochemicals a year—including polyethylene and polypropylene plastics.

All the while, INEOS has kept gobbling up more plants, doubling down again on its plastics bet in 2020 with a $5 billion deal for fourteen more BP facilities. A few years later, it bought 2,300 fracking wells of its own in South Texas. Today, with annual sales of more than $41 billion, and nearly two hundred production sites in twenty-nine countries, it is the world's sixth-largest chemical company, its reach a reflection of the wider industry's long global tentacles. *The Sunday Times* declared Ratcliffe Britain's richest man in 2018, and while he's slipped a few spots since, his net worth is estimated at more than £23 billion (about $31 billion).

"AN IMPATIENT BILLIONAIRE"

At one point, he boasted INEOS was as big as Facebook and McDonald's combined. For a long time, the company and its owner lacked name recognition commensurate with that scale. Both seemed happy to mint money quietly, and INEOS executives used to joke theirs was "the biggest company you've never heard of." It was fitting for an industry whose products, while omnipresent in our daily lives, are little known to many.

But then Ratcliffe embarked on the project that would turn him and his company into, if not household names, then at least familiar presences in British life. In 2018, he put £110 million into an America's Cup sailing team skippered by Ben Ainslie, a towering figure in yachting. The following year, he bought a cycling squad that soon went on—in Team INEOS jerseys—to win the sport's marquee competition, the Tour de France. INEOS became principle partner in a Formula One auto racing team and funded the showy 1.59 Challenge, in which Kenyan runner Eliud Kipchoge, wearing the company's name across his chest, became the first person to complete a marathon in under two hours.

Ratcliffe was just getting started. He picked up two European soccer teams, Lausanne-Sport in Switzerland and France's OGC Nice. "It's pretty special," his brother said of the moment Jim first strode onto Nice's pitch. "Ten thousand fans chanting [his] name. You realize you've bought a pretty significant sporting asset." Soon, he had his eye on an even more tantalizing prize: Manchester United was not only one of the biggest brands in global sports; it was also the team Ratcliffe had rooted for since boyhood. In 2024, he paid £1.25 billion (about $1.6 billion) for a 27.7 percent stake in the club. In an unusual concession, the Glazer family, who held the rest, gave him day-to-day control of soccer operations.

"When a private individual starts hovering over the sports industry with large bags of cash," one sports site observed, "there are a couple of sensible questions to ask." Most notably: Why is he doing it? To

Ratcliffe's critics, the answer is obvious: These investments are greenwash, an effort to buff the image of a company with a gigantic global carbon footprint. Indeed, sports have long been a useful tool for remedying reputational problems, from Virginia Slims cigarettes' sponsorship of women's tennis in the 1970s and '80s to Saudi Arabia's LIV golf tour, part of its wide-ranging effort to paper over human rights abuses. Ratcliffe and those around him pooh-pooh suggestions he's engaged in a publicity ploy, saying his spending stems from personal passion: "Jim is a sports fanatic and he just loves being involved in it," his colleague Crotty said. Either way, it's undoubtedly "throwing an awful lot of positive PR at the company," a longtime chemical industry player noted. "I would love to know who advises [INEOS] on communications," one environmental activist told me with a sad laugh. "Because this is genius."

Brand recognition isn't essential for plastic and petrochemical companies, since they don't sell directly to consumers. But a public profile like the one Ratcliffe has crafted is handy for those looking to befriend politicians. Getting your name in the headlines opens doors. And, the activist explained, "if you have a sports team, you have all these spaces for informal discussions with decision-makers," bringing opportunities to influence those with power over taxes, environmental regulations, and other areas that affect a big multinational's bottom line. "Politicians are just like anyone else, they're just people," Crotty told me. There are few "who haven't heard of Jim now. And that wouldn't have been the case ten years ago."

Taking advantage of the platform his cycling team provided, Ratcliffe used a 2019 BBC sports interview to lash out at British politicians' refusal to allow fracking. Weeks before Keir Starmer—a devoted Arsenal fan—became prime minister in 2024, he joined Ratcliffe at Manchester United's stadium to watch their teams face each other. And across the English Channel, Ratcliffe's new profile would prove useful in advancing yet another ambitious expansion.

"AN IMPATIENT BILLIONAIRE"

ANTWERP IS KNOWN FOR ITS DIAMOND DISTRICT, ITS PLACE AS A FASHION industry hub, and a stunning Gothic cathedral hung with paintings by famous son Peter Paul Rubens. Like many visitors, I arrive in the city—the heart of Flanders, Belgium's northern region—at vast Centraal Station, craning my neck to admire the steel-and-glass ceiling curving high above the trains, then passing beneath a vast arch to gape at an entrance hall whose massive dome echoes that of Rome's Pantheon. The cobblestone lanes in Antwerp's picturesque center are dotted with Belgium's tourist staples—shops selling beer and chocolate, restaurants serving mussels and frites. Few dining there likely realize they are just a dozen kilometers from Europe's largest petrochemical production zone.

Thomas Goorden, a local activist, has promised to show me around that industrial corridor, and after a quick peek at the cathedral—and a stop for frites—I hop on a tram to meet him. After gazing through its windows at Hasidic Jews riding bicycles in their black coats and broad-brimmed hats, I get off in a bustling neighborhood crammed with secondhand shops and Turkish bakeries. Goorden's office is in a bright coworking space where young people on laptops sit at pale wooden tables beneath a rainbow flag. Goorden has a scraggly beard and curly brown hair cut close on the sides and growing long in a mullet at the back. He's in a black hoodie and jeans, and, thanks to a high school exchange year in Kentucky, speaks perfect English. He spends much of his time digging into the complex deals underpinning a plant INEOS is building here—one made possible by Jim Ratcliffe's transatlantic ethane supply line. Known as Project One, it will—if completed—be Europe's first new cracking plant in decades. Goorden and other environmentalists see its construction as a betrayal of the continent's promises to curb its greenhouse gas emissions and cut plastic waste. Their efforts to stop it have turned this charming city into a central battleground in the fight over whether and how the European Union is prepared to back its green words with action.

Goorden had been working on issues like air pollution and urban trees when someone sent him documents about dangerous PFAS chemicals emitted by the American company 3M's Antwerp factory. Tests found high levels of the toxins in soil, groundwater, and residents' bodies, and 3M agreed to stop producing PFAS at the plant. With contamination scandals blowing up in the United States too, the company eventually announced it would cease production of such chemicals worldwide. Goorden took an important lesson from the experience. The work of investigating 3M paid off many times over in harm avoided. "By that rationale, don't invest in planting trees," he concluded. "Invest in suing multinationals."

There are many on his doorstep—ExxonMobil, Dow, and the French oil giant Total are among those with plants in Antwerp. But as Goorden learned more about Project One, it "set off alarm bells all over the place," and he came to believe it had outsize importance, he tells me. There was, of course, the vast amount of plastic its ethylene output would be used to produce. Project One would be an important link in the company's supply chain, cracking imported US ethane into ethylene and feeding that chemical to facilities that turn it into polyethylene, much of which typically becomes single-use items. Goorden was also troubled by the secrecy INEOS's private ownership allowed. Perhaps most importantly, Project One's ethane demand would drive American fracking for decades to come, making it "a big middle finger to climate."

The company sees it differently. INEOS boasts that Project One, with a $4 billion price tag, will be "the greenest cracker in Europe and possibly the world." Onstage at a 2023 business conference, his long frame sprawled across a chair, legs askew and hair mussed, Jim Ratcliffe was asked whether the world really needed more plastic. "I say this a bit glibly, but do you want to live in a cave?" he replied. Modern lifestyles

"AN IMPATIENT BILLIONAIRE"

are "utterly dependent" on the material, in products from contact lenses to cell phones, and people are not going to give them up, he said. "So what you have to do is find ways of doing it in a more environmentally friendly way."

Project One's carbon footprint, he assured his audience, would be half that of Europe's best existing ethane cracker. Its superefficient production would force older, more polluting competitors to close, he predicted, just as modern cars displaced the dirtier models of decades ago. "You don't want to stop that process, because it's a good process for everybody." What's more, he said, while the plant will be powered initially by methane, a fossil fuel, it could eventually switch to hydrogen—a fuel industry touts as a potential climate-saver. "So the day that Belgium can find one hundred thousand tons of hydrogen for us to consume at an economic price, then it becomes carbon neutral." Indeed, INEOS promised Flemish officials Project One would be "climate neutral" within ten years of start-up—but only "insofar as the techniques and infrastructure are available at that time." ("What will they do in ten years if the conditions are not met?" an anti-INEOS group wondered. "Tear the place down again?")

Expanding on Ratcliffe's arguments, Crotty, an INEOS director and spokesman, told me the idea of Project One incentivizing fracking "doesn't make any sense," because if it wasn't buying American ethane, China would snap up the gas instead. And while opponents "have spun it that this is a new plastics plant," he said, its ethylene would replace existing supplies, feeding plastic production that's already happening. "So it's not going to make any more plastics."

Before my trip to Antwerp, I met at a café near my home in London with the environmental lawyer Tatiana Luján, who is leading Project One opponents' court fight, and who happens to live in my neighborhood. Originally from Colombia, she was matter-of-fact and friendly, in a shimmery blue top and round glasses, but there was a hint of sadness in her voice as she dismissed INEOS's claims one by one. Of course a new plant is going to be higher-tech than an aging one, she said. But to her mind, its very existence is antithetical to the idea of a sustainable future, which requires not just phasing out the oldest and most polluting crackers, she said, but refraining from building new ones. In any case, it doesn't necessarily follow that Project One will force older plants to close—more ethylene may just translate into additional plastic production. And even if the company's green claims are to be believed, Luján pointed out, they account for only the operations of the plant itself—not the faraway fracking that will supply its most important ingredient. "By building this ethane cracker, you're locking in that fossil fuel extraction" by ensuring long-term demand. That's true even if INEOS eventually converts to hydrogen, which would only provide energy to run the plant—not displace fracked ethane as its raw material.

Anyway, Luján said, shaking her head, she wasn't buying the company's promises: "I don't trust their public declarations. I trust what I see." She wasn't impressed by the hydrogen claims—like many environmentalists, she sees the fuel as a false solution that shouldn't be used to justify new gas-burning facilities—nor by hopes that climate damage could be prevented by capturing and burying the carbon dioxide Project One emits. "I'm like, 'Show me. Show me the money. Show me that the technology exists. Show me it's possible. Show me it's not just words and empty promises.'"

Luján had been preparing for the possibility of filing a legal challenge to INEOS's plans since 2018, when talk of a new plant began heating up. She'd come to believe the only way to curtail single-use plastics' proliferation was to "go after petrochemicals." In 2019, just after INEOS announced plans for the complex, Pink Floyd musician David Gilmour raised more than $21 million by auctioning dozens of his guitars and donated the proceeds to ClientEarth, the environmental litigation group

Luján works for. When some of the money was allocated to her team, she suddenly had the resources to mount a fight.

At that point, INEOS intended Project One to have two components: an ethane cracker and a plant called a propane dehydrogenation unit, to turn fracked propane into propylene, the main ingredient in the plastic polypropylene. Luján based ClientEarth's first legal challenge on INEOS's failure to conduct the thorough environmental assessment required. A court ordered the company to refrain from clearing trees at its site until it produced a better assessment. Not long after, INEOS postponed—indefinitely, it seemed—its plans for the propylene plant.

Opponents jokingly called the remaining piece "Project One-half." But they weren't laughing when Flanders approved the company's revised permit application. The next phase of Luján's case would focus on nitrogen pollution. The Netherlands—whose border is less than five kilometers from INEOS's site—was in the midst of what many were calling a "nitrogen crisis." For an obscure-sounding problem, it had big political repercussions, and they were spilling into Belgium too. Nitrogen—emitted not only by industrial plants and traffic (as nitrogen dioxide) but also by manure and chemical fertilizers—contaminates air, soil, and water, as well as fueling climate change. There are many large pig and dairy operations in the region, and farmers were bearing the brunt of governments' push to cut emissions. They were furious, and hundreds drove tractors into Antwerp to protest.

Two Dutch provinces were also challenging Project One, and before Luján's case was heard, a court considering one of those claims yanked INEOS's permit. Flanders's failure to pass a law clarifying its nitrogen reduction plans meant there was no guarantee Project One wouldn't worsen the problem, the court said, giving officials six months to address those concerns. Terrified INEOS would decide not to build in Antwerp, the region's politicians quickly hashed out a plan. Farmers saw it as far tougher on them than on industry, and believed INEOS was getting special treatment. "The government wants to push through" a deal "to please" the company, the head of one farmers' group said. "There is no regard for us." They got back on their tractors, driving this time to the Project One site, but didn't get much for the effort. With the new nitrogen law in place, INEOS got its permit in early 2024. Government, a Green Party politician said, was "dancing to the tune of an impatient billionaire."

I'M GLAD I'M NOT THE ONE DRIVING. THOMAS GOORDEN AND I HAVE LEFT his coworking space, and he's behind the wheel on a highway crowded with fast-moving trucks. Many haul shipping containers, likely heading where we're going, toward the port and petrochemical production zone on the outskirts of town. After my visit to Houston's larger petrochemical corridor, Antwerp's—running flat and wide open, alongside the River Scheldt—feels both different and familiar. Sheep graze beneath big wind turbines, and we pass a much older windmill too, with a white brick base and stilled arms, one of the few visible remnants of the villages razed decades ago to make way for factories and shipping yards. There are bike lanes beside the road, and cyclists zip by. Unlike in Texas, there are no schools, homes, or hotels beside industrial complexes. But the giant cylindric and spherical storage tanks, and hulking plants covered with petrochemicals' distinctive spaghetti tangle of pipes, are sights I've seen before.

Goorden pulls over near a sign that says "Project One North Entry," beside several tanker trains parked on a rail line. Construction has begun only recently, and there's little more to see than trailers and cranes, men in bright jackets walking between white tents and heaps of dirt. Adjacent to the site is an older INEOS plant, and when we pass it, I glimpse signs about avoiding pellet loss, beside piles of white sacks likely containing tiny, raw plastic nurdles—like the ones Diane Wilson gathered as evidence for her lawsuit in Texas. It's a hint of the bigger picture Project One is part of. With Dragon Ships docking nearby, the planned cracker would convert their cargo of American ethane into ethylene the company could process into plastic right next door. There are plenty of other potential customers nearby too—as in the Houston petrochemical cluster, plants are crowded together here, one running into the next.

Plastic has been made here for decades. But just before my visit, a private meeting took place that helps explain why more of it may be produced in the years to come. At an Antwerp plant belonging to German chemical behemoth BASF, more than seventy executives from petrochemical and other heavily polluting industries met behind closed doors with European Commission President Ursula von der Leyen—the EU's top political leader—and Belgian Prime Minister Alexander De Croo. "It was very hush-hush," an expert from a lobbying watchdog group told me—"a classic case of privileged access."

Jim Ratcliffe was there, along with executives from fossil fuel and chemical giants, including Shell, ExxonMobil, LyondellBasell, Dow, and DuPont, as well as steel, mining, glass, and aluminum companies. Crotty, the INEOS spokesman, told me he'd encouraged his boss to attend "because his voice would be very loud." Indeed, Ratcliffe got von der Leyen's ear, speaking to her not just with the group but privately too, and "he honestly felt like she was listening. He was quite impressed." Foremost among the complaints companies aired was big polluters' most familiar grievance: Government, they groused, was suffocating them with onerous environmental regulations. "This is the last chance to stop Europe from sleepwalking into off-shoring its industry, jobs, investments and emissions," Ratcliffe warned von der Leyen later, in an open letter. The executives set out their asks in a communiqué they titled "The

Antwerp Declaration." They wanted red tape pared and called for "a new spirit of law-making" that would "stay tuned with industrial reality" and "let entrepreneurship thrive."

It wasn't just rule changes. Companies hoped the EU would cough up cash too. The declaration used the jargon of finance and bureaucracy, demanding "a strong public funding chapter" with "a Clean Tech Deployment Fund for Energy Intensive Industries," and "public de-risking of private investment" in green measures, including support for capital and operating expenses. It boiled down to this: If politicians wanted industry to shrink its carbon footprint, taxpayers would have to pony up.

In particular, companies wanted money for carbon capture and storage—the largely untested, but definitely expensive, technology energy-hungry industries believe will enable them to keep burning fossil fuels. Some envision sending carbon dioxide by pipeline to be buried beneath the sea off Norway. Critics see such schemes as dangerous hype—even if it could be done at scale, capturing carbon dioxide from factory smokestacks itself consumes a great deal of energy, and is typically less effective than promised. Sometimes drilling companies even use the gas to force more oil from underground reserves. Antwerp's petrochemical sector, though, "is betting the bank" on the idea, Thomas Goorden told me, and getting the EU to pay for it is a top priority. "At the end of the day," he says, "who's going to benefit the most from this whole thing is INEOS."

The summit came in the midst of a big political shift in Europe—the same one that checked the ambition of the continent's packaging law. With right-wing parties surging, nervous establishment politicians were backing away from environmental ambitions, and von der Leyen had been watering down a Green Deal climate package she'd once described as "Europe's 'man on the moon' moment." Ratcliffe and his fellow executives echoed that initiative's framing by calling for an "Industrial Deal," and officials keen to boast about big-ticket investments were all ears. When Project One was originally proposed, one opponent told me, it seemed at odds with the continent's plans to reduce both plastic and fossil fuel use. "What's really sad to watch," she said, is that Ratcliffe's arguments about prioritizing corporate needs have since prevailed.

He was well positioned to take advantage of the moment. His sports spending had brought fame to match his wealth, and the transatlantic ethane supply line he'd built made INEOS uniquely ready to launch a petrochemical megaproject. A few years earlier, he "wouldn't have had such a prominent place in the Antwerp summit," the activist told me. Today, he was hobnobbing with Belgium's prime minister and the leader of the EU. "He's a big man now."

It was paying off. Not long before the summit, when Project One's permit was in jeopardy, Ratcliffe visited decision-makers in Antwerp, and De Croo, the prime minister, called in to the meeting. "If we had not seen so much dedication in the Flemish and federal government," INEOS would have pulled the plug on the project, Ratcliffe said after his approval came through. Indeed, an Antwerp official echoed the company's language in an email to me. A "tsunami of goals, rules, requirements and obligations" threatened to shrink the economy, she wrote. "Let us not shoot ourselves in the foot by formulating ever-more ambitious green goals, without taking into account the cumulative impact on our industry." Project One, the official wrote, was an "inspiring example" she was "especially pleased" with.

Guided by such views, Flanders's government gave Project One €500 million ($550 million) in loan guarantees. Britain helped too, with a €700 million guarantee (it came before Starmer took office, unrelated to his soccer outing). "The project was deemed to have potential to cause a number of adverse environmental and social impacts," the UK's export finance agency wrote, according to *The Guardian*. But a "proposed suite of controls," the department shrugged, "should facilitate the management of these impacts." It's evidence Thomas Goorden is right to call

Jim Ratcliffe "an expert in externalizing costs and risks"—that is, foisting his investments' downside potential onto others. It's too soon to know how the Antwerp Declaration's bigger asks will be answered, but the lobbying watchdog I met was worried: "It has a huge amount of momentum behind it."

A few months after my visit, wondering how Project One was progressing, I watched an INEOS video showing construction of the complex's first major component was well underway. The concrete-and-steel storage tank's diameter is the length of a football field, and its circular wall—about twelve stories high—towered over trucks parked inside. When complete, it will be big enough to store eighty Olympic swimming pools' worth of super-chilled ethane, enough to feed the plant for ten days. On the other side of the world, the cracker itself was being made in pieces in China and Thailand, soon to be shipped to Antwerp and, a company official told me, "bolted together like a big Lego set." The opponents' legal challenges were still pending, but with concrete poured—and money sunk in—it's hard to imagine anything stopping INEOS now.

CHAPTER 11

Shipping Plastic, Shifting Blame

*2020—Microplastics are detected for
the first time in human placentas.
473 million metric tons of plastic produced.*

When China closed its doors to foreign plastic waste in 2018, it could have been a moment of reckoning—a firm nudge pushing wealthy countries to look clearly at the mess we were creating and find a different path. Instead, from Los Angeles to Rotterdam to Seoul, those with waste to get rid of simply found new places to send it.

Forty-five minutes outside Surabaya, Indonesia's second-biggest city, Daru Setyorini and Prigi Arisandi watched the resulting wave of plastic crash onto the fields, villages, and waterways around the home where they were raising their three daughters. The couple met as college students in the 1990s. It was a time of political tumult, just before street protests toppled the dictator Suharto, and a bevy of long-suppressed ideas and movements were bubbling up along with the incipient demands for democracy. So the biology study club Daru and Prigi had started together soon evolved from a forum for scientific discussion into a vehicle for defending Indonesia's environment—one that mixed research with activism. They leaned into that combination after graduating, formalizing their work into an organization they called Ecoton. For a long time, they focused mainly on the chemical pollution factories were discharging into the Brantas River, near their home. They analyzed water samples and pushed journalists to cover their findings, determined to call attention to the damage sky-high levels of toxins like mercury were doing to locals' health.

Around 2010, they began to realize industrial discharge wasn't the only threat to the river they loved. Plastic was becoming an increasingly visible problem. Indonesia had long struggled with plastic pollution, largely because many areas lacked formal waste collection, leaving households to dispose of their own garbage. The blight had worsened over the years as plastic packaging grew more common. But what Daru and Prigi noticed as the decade wore on was something new. Suddenly, plastic was everywhere—piled beside roads, littering banana groves, heaped in the middle of villages, and smoldering in smoky fires. With labels in English, German, Italian, and Korean, it was plain to see that much of it was coming from abroad.

Daru sets a tray with a pitcher of guava juice and two glasses onto a long table as she sits down to tell me about it in the courtyard of the compound that serves as her family's home and Ecoton's office. It's a hive of activity—staff members and interns are chatting and laughing, munching on snacks as Daru and Prigi's teen and twentysomething daughters wander in and out. The sound of amplified prayer from neighborhood mosques alternates with dance music from a nearby park to create a nonstop wall of background noise. I'm groggy from jet lag, but also wired with excitement about being here, my first visit to Indonesia in eighteen years. In the 1990s, I spent a year teaching English about three hundred kilometers from where we sit, and I'm basking in the familiar but long-forgotten mix of sounds, smells, and sights. This eastern part of Java, Indonesia's most populous island, is crowded with people and industry, but like Yogyakarta, where I once lived, the area around Ecoton's headquarters nonetheless has a rural, village-y feel. The muddy, fast-flowing Brantas is just across a narrow dirt road, and its banks, at least along this stretch, are thick with vegetation. Banana and tamarind trees dot the garden behind the house, and passionfruit dangles from vines overhead. As I suddenly remember, it can seem like life bursts from every crevice in Indonesia. Now the tropical heat combines with humidity so thick it feels like I've stepped into a bathroom where a hot shower has been running. I didn't need reading glasses last time I was here, and I find the steaminess keeps fogging them up. Smoke from young staffers' clove cigarettes adds a chokingly sweet note to the air, and I take a sip of juice as I soak it all in.

Daru wears a long-sleeved green T-shirt and black leggings, and a dark hijab encircles her round face. With her easy smile and warm manner, it's clear she holds things together here, coordinating Ecoton's many activities, from river tours to microplastics research. She pulls her knees in to balance on a wooden bench as she explains that the twenty or so paper mills nearby recycle used paper they buy from abroad. When China closed to foreign scrap, those paper shipments were suddenly arriving tainted with large amounts of plastic—some the result of poor sorting by overwhelmed recycling companies, some intentionally smuggled by shady operators desperate to offload waste.

Foreign plastic was not new to Indonesia. But as China started restricting waste imports, the volume of plastic landing in the sprawling archipelago nation spiked. Its neighbors Malaysia, Thailand, and the Philippines experienced a similar deluge, as Southeast Asia bore the brunt of the initial wave of trash. Malaysia, by one estimate, took six times more plastic waste in March 2018 than it had a year earlier. In Indonesia, scrap paper imports jumped by 35 percent, even as their reported value rose by just 6 percent—clear evidence, in Ecoton's view, that something else was hidden in those bales of newsprint. Paper mills weren't the only ones getting illicit waste. Shipments that supposedly contained clean, well-sorted plastic bottles for recycling were crammed with an impossible-to-manage mix of everything from sneakers and used diapers to many types of plastic packaging, often soiled with food or other contaminants. "We saw many new dumps," Daru tells me. "More and more plastic."

Over the years, a complex economic ecosystem had grown up around imported waste, and its participants scrambled to scale up. After mills' machinery removed plastic from batches of paper, the companies would sell the plastic into a chain of middlemen. Those brokers in turn passed the waste along to a network of informal sorters, who removed anything with value—wire and other metal, plastic bottles that could fetch a price from recycling firms—and sold it on. That still left vast quantities of low-quality, hard-to-recycle material—bags, pouches, food wrappers, and much more. Workers would shred some and dry it in the sun, then sell it as a cheap form of fuel. Whatever remained would be dumped or burned to make space for the next load.

As the tide of waste grew through 2018 and 2019, paper mill workers confided to Daru and Prigi that plastic accounted for 30 or even 40 percent of the shipments coming in, although an industry spokeswoman told me she did not believe it was ever that high. The couple began looking more closely at the stream of junk. "We tracked the route of the trash—from the paper mills to the dump sites," Daru says. A few of their staff, posing as sorters, bought a load of plastic from a nearby mill. Upstairs from Ecoton's courtyard, Prigi shows me some of what they got. Distressingly, I find much of what is tacked up on a wall display very familiar—bags that once held Doritos, Uncle Ben's rice, Nestlé's Toll House chocolate chips, and the kind of sugar I like to buy from Trader Joe's. After snapping a few pictures, I follow Prigi to an empty office, where he switches on an air conditioner as we sit down to talk.

He's in cargo shorts, sporty sandals, and a black T-shirt, his dark hair floppy on top and buzzed short at the sides. Activism started early for him, he explains, and it was intertwined from the start with his relationship to the Brantas. He grew up along the river, just outside Surabaya, and his great-grandparents used to fish in its tributaries, setting out nets during the night and hauling their catch to market in the morning. As a boy, Prigi swam and played in the water for hours on end, riding the currents with his friends on pieces they cut from banana trees. As he grew, through the 1980s and into the '90s, the paper industry's presence expanded too. So did its environmental footprint. A layer of sludge appeared atop the water, so thick chickens could walk on it, and people had to hold their noses against the smell as they walked nearby. It was part of why, years later, he decided to put the scientific training he had gotten at university to work in the world. He looked at why fish were dying in the Brantas, and went with Daru to study mangrove fields, writing in a newspaper column he'd landed about the threat development posed to those extraordinary trees.

When the wave of foreign plastic hit East Java, Prigi began writing letters to embassies and elected officials from the countries whose waste he was finding—the United States, Canada, Britain, Australia. Through contacts at the ports, he and Daru got documents showing where shipments were coming from. A short documentary they made shows heavy equipment scooping plastic into dump trucks on a paper mill's grounds, trucks spilling huge loads of it onto green fields, and thick smoke rising from heaps burning in the open, as sorters with no protective gear wandered past. One shot shows a US Postal Service priority mail pouch beside a crushed milk carton, and in another scene, Prigi pulls a dirty American flag from a pile, then a torn bag bearing Target's red-and-white logo. The film ends with him going to Jakarta to deliver packages full of waste to foreign embassies. "We don't need this," he tells the camera, tapping an envelope stuffed with junk. "Indonesia is not the developed countries' landfill."

PLASTIC INC.

ALONG WITH PRESSURE FROM ACTIVISTS, INDONESIA'S GOVERNMENT WAS getting heat from around the world, as media coverage of its mess brought an unwanted spotlight. Politicians' initial response proved to be more show than substance. They called press conferences to display shipments purportedly intended for recycling that were stuffed with waste, including rotting food. "We were supposed to get paper scrap, but instead it came with other materials such as plastic bags, rubber, plastic bottles, plastic pouches from cooking oil and soap," an environment ministry official said at one such event in 2019. The government announced it was ordering importing companies to send hundreds of containers back where they'd come from, to countries including the United States, France, Spain, Greece, and Australia. Indonesia's neighbors were taking similar action. In the Philippines, President Rodrigo Duterte insisted Canada take back shipments that, while labeled as recyclable, actually held household trash. "I will advise Canada that your garbage is on the way," Duterte said. "Prepare a grand reception. Eat it if you want to." Such announcements made for good headlines. But after the fanfare died down, it turned out many containers had ended up not back in the wealthy countries where they'd started, but simply sent on to other poor nations, including India, Mexico, and Vietnam.

Still, dragged and prodded by pressure at home and abroad, Indonesia's government also took real steps to stem the flow of junk arriving. Part of what made the country so appealing to shady trash exporters had been its long, painful history of corruption. Cash, it seemed, could often ease a junk-filled container's passage through port. Customs officers "only care about their pocket money," one expert told me. As higher-ups got serious about keeping unwanted waste out, the customs directorate cracked down on bribery in the ranks.

As well as enforcing its rules more assiduously, Indonesia also tightened them. In 2020, the country said it would only accept scrap imports whose impurities—any material other than the main one being shipped—were less than 2 percent. Officials had first pledged a cap of 0.5 percent, but backed down after British and American exporters complained that level would be too hard to achieve. Even 2 percent, though, was a big step. Before, there had been no limit at all.

To ensure compliance with the new rule, officials required every container headed for Indonesia be inspected before shipping. At a stylish pan-Asian restaurant in Surabaya, I meet Lannawati Hendra, an executive at Surabaya Mekabox, which buys cardboard from Europe and Asia to make boxes and paper. She's reserved a room in the back for our lunch, and as is typical for Indonesians, she gives me a warm welcome. It gets only more effusive when she says her sister lives in New Jersey, and I tell her I grew up there. We chat about the sister's experiences before I ask how the new scrap import rules have affected the company. "It's really, really strict," Hendra tells me. Inspectors examine not only every container but every individual bale of paper, before it can be loaded onto an Indonesia-bound ship. Surabaya Mekabox must foot the bill, and the new system has added about 5 percent to the cost of its materials. But the process, she says, is working. Cardboard shipments now contain far less than the maximum 2 percent impurities, she tells me.

Another key piece of the plastic crackdown was a requirement that those seeking to send scrap to Indonesia register with its government. The waste trade is a complex network of brokers buying and selling material that has been tossed into wealthy countries' recycling bins. It's notoriously opaque, and while there are legitimate operators, it is also rife with fly-by-night companies that frequently change their names and addresses. So while it's easy to see from the labels of dumped trash where it originated, the official paper trail can be much harder to follow. Sometimes it's not even clear which country a shipment has come from. At one point, trade data showed Indonesia was getting more plastic scrap from the Marshall Islands, a Pacific nation with a population a tenth that of Minneapolis, than from the United States. Activists believe it was waste from America and other rich countries shipped through a firm registered in the island chain. The new registration requirement was an effort to introduce transparency and accountability into that tangled web.

Indonesia had begun to push back against the flood of waste, deciding, as China did, that it no longer wanted to be a dumping ground for the world's rubbish. As Daru and Prigi show me around their corner of East Java, I can see both the improvement that effort has delivered and the reality that foreign trash is still arriving. It's clear, the couple tell me—and other activists and experts agree—that shipments of recyclables now contain far less unwanted junk. The peak, in retrospect, came in 2019. But the problem, they say, is not solved. Indonesia's experience—a mixed picture of halting progress and continued challenges—is a vivid illustration of just how hard it is to stem a global tide of plastic that grows larger every year.

Just a few years ago, the village of Bangun was notorious as a dumping ground for plastic. It was heaped high in front of homes and spilled across lots beside the little lane that cuts through town. As I head down that road with Daru and Fadiyah—a young journalist who's translating for me on this trip—it looks mostly clear of plastic, although I spot a few small piles. We pull up in front of a house with a tiled front and are soon inside with Rebin, a man in his midfifties, in long shorts and bare feet. Leaning forward on a low blue stool, he tells me he used to buy truckloads of plastic from a nearby paper company and sell them to fellow villagers. He'd get about eight loads a week, and deposit one outside his own house. As his neighbors did with their batches, he'd pull out aluminum cans and plastic drink bottles for recycling companies, burn plastic coating off wire to get at the metal, and dry the rest in the sun to sell to small-scale food producers in need of fuel. While margins were tight, the volume was so great he could eke out a decent living, says Rebin, who like many Indonesians has only one name. "It was enough for us."

But as 2019 turned to 2020—even before the pandemic—paper mills were suddenly refusing to sell waste to informal workers like him. The government, it seemed, had told them to stop. So now, Rebin hauls sacks of shredded bottles for a recycling company, earning about half what he used to, and hires himself out as a farmhand to make up the difference. I ask if he'd like to go back to sorting, and he smiles broadly. "We still hope to do that job again," he says. It's a sharp reminder that the crackdown on waste plastic has yanked livelihoods from some of those who depended on it. Daru tells me later the government hasn't done nearly enough to help them find other work.

After saying goodbye to Rebin, we stop beside an emerald-green rice paddy just outside Bangun and walk a few minutes through stalks of sugarcane more than twice my height. In the distance, a volcano juts skyward, its top hidden in a white cloud, and Daru recalls that this field used to be covered with plastic. It's a heartening transformation, but she believes waste's disappearance from Bangun reflects not only a drop in the amount being smuggled in but also the reality that some plastic is simply being dumped elsewhere instead. A few kilometers away, we drive beside a tributary of the Brantas, motorcycles zooming past, through a village called Gedangrowo. Ripening papayas and stalks heavy with green bananas hang from trees, and fields are covered with piled-up plastic, much of it shredded into small pieces. One such heap sprawls beside the barbed wire–topped wall of a paper mill's grounds, and, as sweat pours off me in the scorching sun, we walk onto the odorless pile (the plastic has likely been washed by the mills' machinery). Acrid black smoke wafts from a fire tended by a man in flip-flops and a baseball cap. Labels suggest much of this waste originated in South Korea, although we also find a Dutch toothpaste tube, a French pack of nuts, and an Australian credit card.

PLASTIC INC.

The next evening, I climb a bit nervously onto the back of a motorbike driven by one of Ecoton's young employees. We zoom past roadside stalls selling fried rice and fried noodles and fried chicken, and as the evening call to prayer echoes from minarets and the whiff of burning trash hangs in the air, I hold on to him for dear life. After a few minutes, we turn onto a narrow lane, and a security guard—my guide seems to know him—waves us into a storage complex where heaps of plastic some six meters high sit behind a large, open-sided shed. The sky is reddening with the tropical sunset, a bright moon is rising, and we're surrounded by the deep green of corn and peanut fields. The scene is eerily beautiful—aside from the piles of waste. I bend down to look at some labels, and find a mix of European, Korean, and Indonesian packaging.

I SEE THE LIMITS OF INDONESIA'S PROGRESS EVEN MORE VIVIDLY THE FOLlowing week, after I've left East Java for the western end of the island. Early one morning, two smiling young women from a research and advocacy group called the Nexus3 Foundation pick me and Fadiyah up at my Jakarta hotel, a small guesthouse in a leafy enclave tucked off one of the capital's many noisy, traffic-clogged thoroughfares. In an air-conditioned SUV, we cruise eighty kilometers along a modern highway—like the capital's sparkling new metro train, it's the result of a major infrastructure push that has transformed the better-off regions of Indonesia in recent years. In a coastal town called Serang, where roadside vendors sell fruit near big factories, we bump down a badly rutted road, and when we pull over, I gape up at a huge mountain of plastic towering overhead. There must be some soil mixed in with the waste, since low bushes sprout near the bottom. We begin climbing through them, and the hill is so high it takes a few minutes to reach the top. Once there, I look around, astonished and overwhelmed by the scale of this pile, struggling to grasp just how big it is. We must be almost a dozen meters up, and among the undulating bumps and rises around me are heaps that tower a further five meters. Hundreds of meters wide and hundreds long, the mountain looks to stretch over an area as big as ten football fields, at least.

As at the much smaller field I saw with Daru, this plastic has been washed—there's no odor—and is mostly shredded. It feels soft and spongy underfoot. On one side, I look down over red rooftops, a mosque's green domes, and the lushness of rice paddies and palm trees. On the other, perhaps a kilometer away, three striped smokestacks sit atop a factory belonging to Indah Kiat Pulp & Paper, one of the nation's largest paper companies—and the source of all this plastic. As I wander around, I spot plenty of recognizable items—a baseball hat, a high-heeled sandal, sneaker insoles, a ziplock bag. I bend down to pick up a packet that once held salt-and-vinegar peanuts in some New Zealand shop—"we've been bringing nuts to Kiwis for over 30 years" its label announces—and a bag of freeze-dried strawberries from Australia. There's plenty of American waste—hot dog wrapping, an instant clam chowder pouch, scraps emblazoned with the name of Costco's store brand, Kirkland Signature. Trader Joe's is no bigger a contributor than any other chain, but it's my favorite place to shop when I'm back home in the US, so its name jumps out at me, and I spot packages of its roasted chicken breast and vegetable fried rice. Before we descend, I examine a piece of laminated paper that bears an image of the Virgin Mary on one side and a woman's photo, with dates of birth and death, on the other. She was 108 when she passed in 2019. I can't tell what country this funeral home memorial card comes from, but it's in English. Even in death, those of us in the developed world leave a trail of trash behind.

We descend and cross the street, then climb a low rise to a small shack on a lot where plastic is piled all around. A woman in a dark pink blouse and flowing trousers tells me in English that that this is a sorting business her family owns. The paper company adds to the mountain every day, and scavengers climb it to bring her material to resell, she says.

A group of those workers are sitting on a low table beneath a torn blue tarp, and Fadiyah and I approach a woman with big, dark eyes and dirty bare feet. Her name is Kasih, she tells us, and she climbs the plastic mountain every day, after she finishes her morning job, selling bananas. I pick up Indonesian words here and there as Fadiyah translates, and Kasih says the packaging discarded in faraway countries is hot to the touch when she digs through it. She works with her husband, who sits beside her now in rolled-up jeans and a bright orange top. Carrying what they find in big sacks, they earn $2 to $4.50 between them from seven hours' work. "It's very exhausting" and sometimes leaves her struggling for breath, Kasih says. The money is just enough to cover the cost of fuel and rice. "As you can see, I'm just a poor person," she tells us, her voice clear and strong as she holds my gaze.

Later, when I contact the paper company, Indah Kiat, I expect its representative to deny the mill is responsible for the vast plastic mountain. But the spokeswoman I reach acknowledges the waste arrived in the company's used paper shipments. Despite the heap's scale, she tells me Indah Kiat's imports meet the government's 2 percent limit on impurities, adding that the company adheres to all regulations, and is shifting toward burning its waste plastic as fuel. Similarly, the head of the association representing the Indonesian paper industry says all its member companies' imports comply with the 2 percent rule. Environmentalists believe contamination, while much reduced, often still exceeds that cap. Either way, it's clear that what's arriving still adds up to an extraordinary amount of plastic.

IN PRINCIPLE, OF COURSE, MUCH OF THE PLASTIC COMING TO INDONESIA is intended for recycling. But like everything I've learned so far about this country's relationship with imported waste, the reality is complicated. Indonesian environmentalists argue that, given the pollution the process creates, if wealthy countries want to recycle plastic, they should do it at home. "How come it's our problem?" one activist asked me. "It's your mess. You should be able to help yourself."

Still, there is a market in Indonesia for well-sorted plastic scrap, from companies that turn it into material they sell to manufacturers. Accompanied by Daru and Fadiyah, I visit one such business not far from Ecoton's office. We enter Pelita Mekar Semesta's plant through a fenced-in lot covered with big bales of plastic, then head upstairs to a conference room with exposed-brick walls and big windows that overlook a clean, modern factory floor. Chresna Agung, a company vice president, has prepared a slide presentation for me. The plant imports waste plastic from abroad, I learn—the Netherlands is its biggest overseas supplier at the time of my visit—and sources domestic material via middlemen who buy it from small-scale scavengers and sorters like the ones I've met. The company processes a variety of plastics—bags, jugs, yogurt tubs, hard material like that used for buckets—into eighteen thousand tons of pellets and 4,800 tons of plastic film each year, which it sells to manufacturers at home and overseas. Jars of sample pellets and rolls of bag-like material, in a rainbow of hues, sit on the table in front of me, and in a corner of the room I see brightly colored pails, crates, wastebaskets, hangers, and toys, all made from recycled plastic.

Agung talks me through the process—sorting, washing, drying, pelletizing—but it makes more sense when I see it. We don helmets and masks, and start our tour in a giant warehouse, where loud music plays and several hundred women, in lavender company shirts and brightly colored headscarves, hijabs, and skirts chat as they sort through bales of clear plastic wrapping. We walk through a series of rooms—the heat is oppressive and the noise earsplitting—where sorted plastic is chopped, washed, and dried, and eventually we reach the spot where pea-size pellets pour into big white sacks. Agung puts a few in my hand, but they're so hot I drop them after a few seconds. In another room, sheets of plastic film, in varying widths and colors, roll off big machines, ready to be turned into trash bags, shopping bags, and more. Plants like this are still a rarity in Indonesia, but it's an impressive facility, and as we walk out the front gate, past people on motorbikes waiting to pick up spouses coming off their shifts, I think about the value of turning used plastic into new. A bandage on a gaping wound, yes—but perhaps a bandage we need.

Not all Indonesian recycling is so professional. By the side of a road in an East Java village called Gedek, I'm hit by the smell of burning plastic and the roar and rattle of ancient machinery as I walk into an open-sided space beneath a corrugated metal roof. Three young men—two might be in their teens—tend a Rube Goldberg sequence of rusting equipment. Sachets—tiny packets, like those that hold ketchup, mustard, and soy sauce in rich countries, but which are used in the developing world to sell a much wider range of products—are heaped on the ground. Two of the workers stuff handfuls through a smoking metal funnel into a wide, horizontal pipe. It takes me a while to realize that the goopy brown sludge oozing from the other end is melted plastic. Another machine molds it into strings, and long strips like dark spaghetti run through a metal bathtub to cool, then over a straw mat to dry, before being chopped into pellets. Every few minutes there's a loud bang. It couldn't be further from the gauzy green image of recycling we've been sold, but this is the reality for some of the lowest-quality plastics. The waste here today is mostly Indonesian, but Daru tells me the local dealers who supply this operation buy foreign plastic from paper mills too.

Indonesia, of course, isn't the only country struggling to manage the junk coming from wealthy nations. After my trip, I speak with Jim Puckett, who runs the Basel Action Network, a watchdog group that monitors the waste trade. He peppers his conversation with technical jargon I have to look up later—like all the most effective environmental activists I've met, he's steeped in the regulatory intricacies of his field. His group's

SHIPPING PLASTIC, SHIFTING BLAME

name is a reference to the Basel Convention, a 1989 treaty that set rules for developed nations sending hazardous waste to poorer ones. In 2019, delegates from 187 countries amended it to add the first-ever global rules on cross-border shipments of plastic. No longer could exporters ship contaminated, mixed, or unrecyclable plastics without the recipient nation's informed consent. It was a landmark moment, and Puckett hoped the new agreement would be a big step toward addressing the problems laid bare by China's closure.

Reality has not lived up to that ambition, he tells me. The plastic amendments' promise was limited from the start by the absence of the United States, which hasn't ratified the Basel treaty. Both it and the convention's signatories are punching loopholes into the new rules with egregious interpretations of the amendments' language, Puckett says. Rich nations "are finding ways to wiggle out from under the agreement," and poorer ones "are just going, 'Well, we're not going to bother.'"

And while Indonesia is among the nations beginning to get a grip on imports, the scrap trade's global web is an ever-shifting game of cat and mouse. When one country erects barriers, those with material to get rid of often just find someplace else to send it. The US shipped a lot of waste to Southeast Asia right after China stopped accepting it, for example, but has more recently shifted exports toward Latin America, particularly Mexico. European nations that once favored Malaysia later began sending more to Turkey. Despite that disheartening reality, it's also true that plastic scrap exports overall have been declining since 2018. Wealthy countries seem to be landfilling and incinerating more of their plastic at home. That counts as a win, of sorts, if not for those who thought the stuff they threw into their recycling bins would actually be recycled, at least for the poorer countries getting less of it.

Fundamentally, Puckett tells me, shipping waste in any form is about pushing the costs of dealing with it onto someone else. That, in his view, applies not just to the stuff sent illicitly—like plastic smuggled inside bales of paper—but also to legally shipped material that ends up at modern recycling facilities like the one I visited. "The word *recycling* always has this green aura about it," he says, but with plastic, that's deceptive. Minimizing the process's harms is expensive, as is the laborious sorting and cleaning required. Exporters profit from off-loading such expenses. Importers gain by cherry-picking the best material and dumping the rest. Margins may be low, but volume is huge, so the profits add up. "This is how you make money off of waste," Puckett says. "You don't have to pay for the stuff you should be paying for."

OF COURSE, NOT ALL THE PLASTIC CHOKING INDONESIA IS RICH COUN-tries' waste. Indonesians themselves are using a growing amount of disposable packaging too. That's no accident. Emerging markets are key to industry's plans to make and sell ever-increasing amounts of plastic, and companies have been pushing it aggressively across the Global South. Unlike in wealthy countries already saturated with plastic, there is plenty of room to sell more in poorer nations.

It wasn't always this way. During my year as a Yogyakarta English teacher, in the early 1990s, my American housemate and I often shopped at a small neighborhood food stall. We had to laugh the day a purchase came wrapped in a draft of the graduate school application essay she'd been working on. Someone was mining our trash for paper. That ethos of reuse was once common. When Daru was a schoolgirl in the 1980s and '90s, she tells me, she and her mother bought vegetables bundled in newspaper, and brought their own baskets, jars, and jerricans to the market. Even as the millennium turned and plastic packaging grew ubiquitous, she says, "it was bad, but not as bad as now."

The shift was helped along by a relentless barrage of advertising portraying food, drinks, and household products packaged in plastic as modern, hygienic, and high quality. At the same time, companies ramped up their use of plastic packaging and eliminated alternatives. As the years went by, tea once sold in glass bottles was only available in plastic, and drink mixes that had long been sold in tins appeared on shelves in pouches instead. "People have no choice," Daru says. "They have to buy plastic."

A particularly pernicious form is sachets, single-serving packets like those I saw melted down at the roadside recycler. Across developing Asia and Africa, big food and consumer products companies sell a huge range of products in them, everything from shampoo and laundry detergent to coffee and spices. Such use began in India, where poverty had long motivated small purchases: In the 1970s rural shopkeepers would sometimes carve a single bar of soap into a half dozen pieces. Late that decade, in the southern state of Tamil Nadu, Chinni Krishnan, a farmer with a sideline in pharmaceuticals, saw from children's matted hair that many families had no shampoo, and he vowed to give India's poorest access to modern goods. "Whatever I make, I want the coolies and the rickshaw pullers to use," he would tell his three sons. "I want to make my products affordable to them." Krishnan started repacking talcum powder and Epsom salts into small quantities, and experimenting with ways to do the same for liquids. He soon hit on a method for sealing up little plastic packets, and in 1976 began using them to sell a shampoo he called Velvette, as well as hair oil and honey. "This is going to be the product of the future," he told his boys.

Krishnan was better at invention than promotion. It was his youngest son, C. K. Ranganathan, who brought sachets to the mass market, after the patriarch's 1979 death. Splitting with his brothers to form his own company, he named his shampoo Chik, in honor of their father, and sold it in sachets for three-quarters of a rupee each, about six US cents. Ranganathan had the business gene his father lacked, and a knack for reaching the rural poor. He staged live demonstrations to show villagers how to use shampoo, inviting onlookers to feel and smell young boys' hair after it had been washed. Consumers didn't see shampoo as an essential, but his tiny packages' low price made it less risky to try. He was good at distribution too—he put Chik on sale at weekly markets where many did much of their spending, and placed ads at movie screenings, offering free samples after the show. Before long, Chik was southern India's top shampoo brand, and as it gained popularity, Ranganathan began selling other products in single servings too, from detergents to cooking sauces.

At first, his big competitors hadn't taken much notice. The Indian arms of giant multinationals weren't interested in low-income customers, preferring to sell standard sizes to wealthier Indians. "They came into the sachet business a decade after us," Ranganathan later recalled. But by the end of the 1980s, Hindustan Lever Limited, Unilever's Indian subsidiary (now Hindustan Unilever), jumped aggressively into sachets. "We discovered that wealth lies in rural India, and we reached out to the wider market base," the subsidiary's then-chairman later said. Competitors such as Procter & Gamble and Nestlé followed suit, not just in India—where by 2000 nearly 70 percent of shampoo was sold in sachets—but across the developing world.

What began as a way to give the poorest access to modern goods has exploded into a booming global business. Filipinos, for example, use 164 million sachets a day. Annually, that's almost six hundred per person, or about sixty billion in total—enough to bury the capital, Manila, thirty centimeters deep, by one estimate. Worldwide, just over a trillion sachets were produced in 2023, and annual output is expected to climb past 1.4 trillion in a decade. In many countries, more shampoo is sold in sachets than in large, multi-serving sizes, and that holds true for a number of other products too. The market for sachets—just the packets, not the goods they contain—is worth nearly $12 billion a year and is expected to surpass $17 billion by 2033. "Sachet packaging has a bright future," one analysis concluded.

That future has a darker side, though. Because of their size and composition—layers of different kinds of plastic, typically—sachets are very hard to recycle. At best, they get melted down and combined with other plastics to make twine, planks, and bricks. That processing releases toxins, and the new goods eventually degrade and release microplastics. And because sachets are all but worthless on recycling markets, there's little incentive for informal sorters to collect them, so they accumulate in dumps, litter landscapes, and clog rivers. Often, they're simply burned.

Even an executive from Unilever, long one of the world's biggest sachet users, declared in 2019 that the sachet is "evil, because you cannot recycle it." The following year, the company's then-CEO acknowledged "we have to get rid of them" for that reason. Nonetheless, a Reuters investigation found the consumer products behemoth lobbied against proposed sachet bans in Sri Lanka, India, and the Philippines. "Unilever approached us and said: 'Don't do this, sachets are a poor man's commodity,'" Sri Lanka's environment secretary told Reuters. "We said: 'Yes, you have addicted the poor man to sachets. Now they have no choice.'"

I get a glimpse of sachets' role in Indonesia when I meet Daru at dawn one morning in Ecoton's courtyard, roosters crowing nearby, to tag along on a shopping trip. Modern supermarkets and convenience stores are common in Indonesia, but there are still many traditional markets where vendors sell produce, fish, meat, and other items. We walk along the Brantas for a few minutes, past women sweeping front porches, to a little dock where we step onto a flat boat. Using a long pole to push off the river's bottom, its pilot delivers us swiftly to the other side. While Daru selects ingredients for her family's dinner, I wander aisles where bright pink dragon fruit and whole coconuts are heaped on tables, and tall stalks of bananas lean against wooden counters. Much of the produce is unpackaged, and some shoppers have their own bags and baskets. But above the enticing piles of fruit, shiny, colorful packets hang in long strips—flavored drink mixes, powdered coconut milk, hair products, fried chicken seasoning, a multitude of sauces. "Everything is packed in sachets," Daru sighs.

And it's not just sachets. Packaging use overall is increasing by 4 to 6 percent a year in Indonesia. Flexible plastic packaging—thin, pliable material like that used in pouches, wrappers, and bags, as well as sachets—is growing even faster, by at least 8 percent annually, says Ariana Susanti, director of the Indonesian Packaging Federation, which represents companies that make and use packaging. To her mind, Indonesia's plastic pollution results not from that relentless growth, but from individuals' failure to dispose of and sort their trash properly. "We need to educate people" to stop littering and bring recyclables to collection sites, she tells me. For their part, she says, companies are working on redesigns, trying to construct pouches from a single type of plastic rather than many, to enable better recycling. But there's little industry can do to stop sachets' proliferation as long as consumers continue to prize their convenience. "People want them," she says.

In Dini Trisyanti's view, there's nothing wrong with that. A waste consultant who works for big global brands, she's fed up with a public discourse that paints plastic as evil, she tells me over lunch on the top floor of one of Jakarta's many high-end shopping malls. Environmentalists fan hatred of the material, schools brainwash kids against it, and trend-conscious Instagrammers fetishize metal water bottles, says Trisyanti, a small, friendly woman in orange sneakers and a headscarf secured with a sparkly pineapple pin. "I don't see plastic as more sinful" than other materials, she tells me, as I alternate between stuffing sweet, spicy morsels of lunch into my mouth and taking notes on my laptop. She believes the fixation with plastic is a distraction from Indonesia's real problem, a grossly inadequate waste management system in which only 65 percent of trash is collected. Wealthy countries, she points out, use far more plastic per capita than Indonesia, but less ends up in the environment because they dispose of it properly. That is certainly true. But like

Susanti's emphasis on individual responsibility, Trisyanti's argument—that plastic usage can grow unchecked as long as governments handle the waste—echoes the framing companies have long favored.

Daru, of course, sees things differently. People opt for single-use plastics not only because they're convenient, she says, but because they're cheap. That low price, though, doesn't reflect their true cost—not the monetary cost of managing waste, nor the environmental and health damage it wreaks. Companies "want us to use more and more and more plastic," and anything other than reversing that endless growth is, to her mind, a false solution. "We need to go back to that era when people bring their own bag to the market" and vendors refill containers—"the old way of shopping" she remembers from her youth. "We can't solve the plastic problem only by treating the waste. We have to stop using plastic as much as we can," she says. That, she knows, means stopping industry "from flooding us with these plastic products."

AS INDONESIA WAS SUFFERING THROUGH THE WORST OF THE PLASTIC deluge in 2019, Donald Trump, then in his first presidential term, made a visit to London. At a news conference, he was asked about climate change, and something caught my ear as I watched video of him rambling around the general idea of environment. "I also see what's happening with our oceans, where certain countries are dumping unlimited loads of things in it," Trump said. "They float, they tend to float toward the United States." He'd regurgitated—incoherently, yet in a way that captured its essence—an idea others were articulating more clearly.

As it happened, I knew what he was alluding to. Over the late 2010s, as images of oceanic plastic pollution mounted into a major public relations crisis, the industries that make and use plastic found a convenient way of talking about the mess. Marine plastics, they acknowledged, were indeed a problem. Not a sign that we should use less plastic, but an indication that the world was not managing it properly. It was the old, familiar littering framing—writ large, this time, on a global scale. In this new iteration, the litterbugs were a handful of developing countries whose inadequate trash collection left vast amounts of plastic entering their rivers, and ultimately the seas. The argument drew on a 2015 study in the prestigious journal *Science* that found five Asian nations—China, Indonesia, the Philippines, Vietnam, and Sri Lanka—accounted for more than half the plastic entering the oceans from land.

Trump didn't say any of that on his visit to London. He didn't even mention the word *plastic*. But his assertion was a clear echo of the idea industry was pushing. He'd added his own twist, weighting a corporate effort to offload responsibility for an environmental disaster with something darker: the idea that foreigners and their trash were threatening America.

The companies' version was gentler, and delivered with more finesse. In 2019, the American Chemistry Council helped create the Alliance to End Plastic Waste, bringing together fossil fuel and petrochemical giants including ExxonMobil, Shell, Dow, and Saudi Arabia's SABIC with packaging users like PepsiCo, all eager to demonstrate their concern. It was another chapter in these companies' long history of using green-sounding language to co-opt—and, they hoped, defuse—anger about pollution. "Everyone agrees that plastic waste does not belong in our oceans or anywhere in the environment," said the group's first chairman, Procter & Gamble's then–chief executive David Taylor. The alliance used its messaging muscle to push the idea of ocean plastic as a result of not excessive production, but the waste-handling failures of a few nations. "Sixty percent of plastic waste in the ocean can be sourced to five countries in Southeast Asia," the group said in its first statement. To further underline where it believed the roots of the problem lay, the alliance based itself in Singapore.

The framing enabled industry not only to appear worried about plastic pollution but also to present itself as a benefactor ready to help fix it. The Alliance to End Plastic Waste promised $1.5 billion for projects focusing around four "pillars" (none, of course, involved selling less plastic): expanding waste management infrastructure, improving recycling, cleaning waste-strewn areas like rivers, and promoting "education and engagement" to "mobilize action." The work would focus "where the challenge is greatest," mainly Southeast Asia. The programs sounded lovely—helping with Sri Lankan coastal cleanup, reducing waste leakage in Bali—but many were tiny, ill conceived, or fizzled after well-publicized launches, a *Bloomberg* investigation found. Four years after its creation, the alliance reported having "diverted" from the environment only 0.2 percent of the fifteen million tons of plastic it had promised to remove over five years, *Bloomberg* said. And much of the promised $1.5 billion turned out to be money member companies had planned to spend anyway.

FOR SOUTHEAST ASIA, THE INDUSTRY'S FRAMING ADDED INSULT TO INJURY. It's bad enough that Indonesia and its neighbors have been hit by a plastic double whammy—waste shipped from abroad, and the packaging huge global companies are pushing into their markets. Now they're getting blamed for the global environmental mess that has resulted from companies' long love affair with plastic.

Jenna Jambeck never imagined her research would be used to underpin that kind of finger-pointing. A University of Georgia environmental engineering professor, she works to trace plastic's pathways through the environment—counting and analyzing pieces of trash from the Ganges to the Mississippi, helping cities and countries understand where their plastic waste comes from and where it ends up, and collaborating on the creation of a debris-tracking app. In 2022, she won a MacArthur "genius grant."

Jambeck was the lead author of the 2015 *Science* paper industry cites in its blame-shifting. But she and her colleagues had not set out to pin responsibility on specific countries. Their project was an effort to quantify the amount of plastic entering oceans. Indeed, the number they came up with—eight million metric tons annually—was far greater than previously thought. It was painstaking work. They'd analyzed waste statistics for 192 nations, combining them to get the global figure. They knew it was just a first stab, and urged scientists to dive deeper into their own countries' data, to refine the estimates. So everyone could see how they'd reached their total, they published the nation-by-nation breakdown.

"It was my desire to be very transparent," Jambeck tells me when we speak on Zoom, and to "make the data as open as possible." In retrospect, she says, what happened next shouldn't have been surprising. But she didn't see it coming. The paper drew a blizzard of attention—even inspiring an *Onion* spoof—and, to be sure, much of the coverage highlighted the stunning global total. But some headlines emphasized the outsize share of ocean plastics coming from the top five countries. And that, of course, was what industry seized on.

Jambeck chooses her words carefully when I ask about the way plastic companies used her study. It's a scientist's precision, but also reflects a hard-won awareness that what she says—and writes—can sometimes be taken in ways she didn't intend. Transparency is a key scientific value, and she still thinks publishing the country-by-country breakdown was the right thing to do. But "a challenge related to being so open and transparent is that then the data can be used by others for their own needs," she tells me. "That means there are people that have different agendas" citing work in ways its authors can't control. Companies making plastic "don't want to decrease" output, and it's clear to her their focus on a few countries' waste-handling failures was a way of shifting attention from their own lucrative business model, she said.

The paper also landed with a thud in Indonesia, where many were distressed to see their country listed as the second-biggest source of marine plastics. Jambeck's name comes up a lot with the people I meet there—"the famous report," Trisyanti, the industry consultant, called her paper—and depending on who I talk to, it's cited as either a needed catalyst to action or an unfair slur. An environment ministry official I meet mentions Jambeck right away. In his view, her team's paper was inaccurate, but he says it nonetheless was "good for alarming Indonesia," helping prompt plans to improve trash collection and reduce excessive packaging.

As for Jambeck, she and her colleagues heeded their own call for further exploration of individual countries' waste figures, launching a new project to more deeply analyze the US contribution to ocean plastics. The original paper had not accounted for waste exports, so US plastic dumped in Indonesia counted toward Indonesia's total, not America's. That changed in the new study, published in 2020. The team, led this time by Jambeck's colleague Kara Lavender Law, also looked more closely at the small sliver of US waste being littered or improperly dumped domestically. It was no shock to find America was the world's biggest generator of plastic waste. But because US trash management is well developed, it was more surprising that about a million metric tons a year—a small fraction of the enormous amount of plastic Americans discard—was ending up in the environment. And when that amount was combined with the chunk of its exported waste that eventually reached oceans, the United States jumped from the twentieth-biggest source of marine plastics to—at the upper bounds of the new estimate—third. Indonesia, with its own trash management problems, climbed from second to first. But with plastic from Europe, Australia, Japan, and other wealthy parts of the world still landing there too, it turned out much of the rubbish Indonesia and its neighbors were getting blamed for wasn't theirs after all. It was ours.

Epilogue

It's painful to look clearly at the ways huge, extraordinarily profitable companies have foisted plastic on us for decades and convinced us it's our fault, while wielding their money and power to squelch any efforts to stop them. But having stared that reality in the face—understanding, at last, how we've gotten into this mess—we are free now to ask what a different relationship with this material could look like. There are no simple answers to the questions posed by plastic's spread into every corner of our lives. It has helped shaped modernity, and it is not going away.

Nor should it. But as I've examined the forces that have shaped our plastic-choked present, and gazed ahead at the path—strewn with even more of it—that industry hopes to lead us down, I've come to understand there are other roads we could take. Reversing plastic's relentless proliferation, and the harms it is wreaking, won't be easy, but it's not impossible either. The urgency of doing so grows every day, along with the volume of the stuff accumulating all around us.

This book has not been about providing a bullet-point list of policy proposals. By profession and disposition, I'm more inclined to probe a problem's causes than detail prescriptions for fixing it. But at the end of a journey both sobering and liberating, it feels important to imagine how we might find our way to a different, less plastic future. To look for the hints of progress popping up even as the tide keeps flowing in the other direction. And, given the lengths to which industry has gone to confuse us with greenwash and false solutions, to untangle what's real from what's not.

The past holds some clues, and we don't need to look far back to find them. Just a couple of generations ago, reusing and valuing objects was more common than tossing them away. Even a decade or two back, with disposability not as utterly dominant as today, single-use plastics were less pervasive. Of course we can't just turn back the clock, and we've reaped benefits—in comfort and convenience—from the shift, even as we worry over its excesses.

Fundamentally, it is plastic's low cost that has enabled its spread into every crevice of the modern economy. That is what makes possible the stream of more always waiting to be briefly used and then thrown away. It's why we so often find ourselves with packaging we didn't want or ask for. Why restaurants sometimes prefer throwaway dishes to the hassle and expense of washing real ones, and why building take-out systems based on reusable containers has never made economic sense. But what I've come to understand, and what we must not lose sight of, is that plastic's rock-bottom price results from companies' success at pushing its true costs onto others. Those costs are paid in damage to our health, our climate, and the natural world, as well as in the dollars and cents, pounds and pence, and all the other currencies with which local governments and taxpayers everywhere must manage the waste plastic inevitably becomes.

Most of those studying how to stem plastic's spread believe changing that equation, by handing at least some of those costs back to compa-

EPILOGUE

nies, is key to ending the most egregious waste. The idea, known in experts' lingo as extended producer responsibility, is, in essence, the "polluter pays" principle that has guided environmental progress in areas from smog to toxic dumping—the belief that the one who made the mess must foot the bill for cleaning it up. "There are people that have gotten very wealthy" off a business model based on avoiding that bill, a California state senator named Ben Allen told me, perched at his kitchen counter in an LA Dodgers cap when we spoke on Zoom. He spent years shepherding a groundbreaking producer-responsibility bill into law in his state. Passed in 2022, it gave companies a decade to cut disposable plastic packaging use by a quarter, and required manufacturers to fund the recycling systems previously paid for by local governments. They must also contribute to a $5 billion fund to address plastic pollution's harms to health and the environment—an effort to begin rectifying the damage already done.

Maine, Oregon, Minnesota, and Colorado have also enacted producer responsibility laws. In 2025, New York state lawmakers declined to bring one to a vote for the second year running. Advocates estimated that if passed, it could have generated $150 million annually for New York City alone. More than fifty-five years earlier, at the industry-dominated packaging waste conference we peeked in on near the start of this book, one of the city's sanitation officials suggested such a fee be levied on packaging producers. The years lost since then—and the problem's exponential growth as they rolled by—are a measure of the industry's success at using its influence to forestall action that might have hurt its bottom line.

Elsewhere, in varying forms, producers' responsibility for plastic waste is already a reality. Germany has been a leader, and other European countries have embraced such laws too; many are now being strengthened as the EU pushes its broader effort to reduce plastic packaging. Britain began in 2022 to charge a tax on plastic packaging containing less than 30 percent recycled material. It also approved a producer-responsibility law that would funnel payments from firms making and using packaging toward local authorities' waste costs, although it's delayed the start date for collecting them.

The power of these laws comes not only from their easing of municipalities' financial burden but also the way they reset incentives, pushing companies to produce less waste instead of more, Judith Enck told me. She's a former US Environmental Protection Agency official who has become one of the plastic and petrochemical industry's most trenchant critics. When I met her one sweltering morning on the back porch of her home in the woods outside Albany, New York, she pulled out a box of smart products and packaging—paper candy bags that lack the usual plastic coating so are fully recyclable; shampoo sold as a bar instead of in a bottle; a glass soap pump that can be refilled by mixing tablets with water. "This is not rocket science," she said. A strong producer-responsibility law, she believes, could move such items from niche to mainstream—from green eco-stores to ordinary supermarkets. "The only way to change the trajectory is with strong laws," Enck told me. "That's my theory of change. You have to pass a law."

Long unachievable nationally in the United States, such a law is unimaginable now, under an administration making polluting industries' wildest dreams come true, and smashing the institutions that have delivered decades of progress on other environmental threats. Plastic, though, is an issue on which change, in America and around the world, has often bubbled up from below, with action from local, regional, and state governments.

Industry has often been successful at stopping such efforts. But not always. So the advocates keep pushing. Enck's group, Beyond Plastics, trains concerned individuals—people trying to use less plastic in their own lives—to press for change in their town, city, or state. "That's some of the most important work we do, is introducing people to the legislative process and urging them to jump in," she told me. Most "are really interested. They want information and they want to know what they can do." She encourages them to start by urging their city council to enact what she calls "the plastic trifecta"—a law prohibiting polystyrene (Styrofoam) food packaging and plastic bags (with a fee on paper ones) and requiring plastic straws be given only on request. "Those bans have been passed in many communities and a couple states," she said.

Of course, narrow bans on specific items are small steps in the face of a very large problem. But they're a place to start, something to build on. And, as industry understands better than anyone, they can ripple more widely than we imagine—showing change is achievable, triggering copycat measures, and beginning to shift markets. There are stories of success across the country, and around the world. In Hawaii, after years of fruitless fights at the statehouse, a band of activists pushing for a ban on single-use plastics for take-out food shifted their focus to the Honolulu City Council instead. At the state level, one activist told me, there are "so many people that have to say yes, so many chances for it to die," but enacting a city law was simpler. A teenage activist who'd grown frustrated joining repeated beach cleanups only to see plastic return with the incoming tide collected signatures from 1,500 young people and displayed them on an eighteen-meter scroll at the council hearing. The bill's backers enlisted support from local business owners, urged concerned citizens to submit testimony, and emphasized the cost of dealing with waste along with its environmental harms. Not long after Honolulu enacted its ordinance, nearby Maui County followed suit. "If you can organize somewhere and pass something," a state lawmaker told me, "you create a foothold. And if you do that across the country, in different states and cities and counties," then the conditions for further progress are in place.

In Europe, where continent-wide change has been more achievable, some cities, and even countries, are pushing ahead of their neighbors. France requires reusable dishware, cups, and utensils for eating in at restaurants, including fast-food establishments. Tübingen, a university city in southwest Germany, introduced a tax on single-use plastic take-out containers and cutlery. It's survived (so far) a legal challenge from McDonald's and is intended to encourage the growth of companies providing reusable options. Such an approach is hard to envision from where we stand today, accustomed to take-away systems built on disposability. But it's not unimaginable, and all across the economy, in sectors from tech to pharmaceuticals, corporations have done far more complicated things. "If you don't sit down and try to figure it out, how are you ever going to change it?" one California politician asked me. The existing system "is the product of many forces, and there's nothing inherently right about it," she said. "The status quo has a lot of power because everything is built around it, and people are invested in it," but that doesn't mean it can't change.

Indeed, in Europe and North America, start-ups have already popped up to offer reusable take-out boxes, bowls, and cups that can be shared among restaurants. Customers can opt for reusable packaging when ordering, then return containers to a delivery person on a subsequent drop-off, or leave them at a participating eatery. Such companies are beginning to work with existing delivery services like Uber Eats and DoorDash. Big events are another place where reuse could replace disposability. In Britain, vendors at Wimbledon and some Premier League soccer stadiums have begun selling drinks in returnable cups that customers leave in designated bins or bring back to reclaim a deposit. In 2024, the Paris Olympics did the same, and music festivals have begun using them too, in collaboration with companies that provide the cups and mobile washing stations; smaller venues like theaters are also embracing reusables. In Åarhus, Denmark's second-largest city, dozens of cafés have joined a program that lets customers opt for a reusable cup, then return it at an on-street bin that scans it and refunds a deposit digitally. Cups were returned three-quarters of a million times in the first year. Elsewhere, in Seattle and the Bay Area, Sydney and Melbourne, London and Amsterdam, similar efforts are underway on a smaller scale.

It's not just takeout, of course. In parts of Europe, stores sell some goods in reusable containers that shoppers bring back on their next visit, and even big brands have been dipping their toes in, with trials of, for example, chocolates in returnable tubs. The EU's new packaging regulation requires larger markets to set aside 10 percent of their floor space for refill stations. "It exists already," one activist told me. "Now what we're going to see, hopefully, is a scale-up."

If such changes are to take root and grow, governments and businesses must work together building the necessary infrastructure. Laws and rules requiring at least some reuse would nudge the process forward. And, as the European activist reminded me, "reuse is cheaper if single-use plastic actually pays the fair price." It might not be easy, it will never be perfect, and there are sure to be glitches along the way. But it's doable. And there could be profit in it too: The World Economic Forum says reusable packaging is a $10 billion opportunity.

In Jakarta, Indonesia, I met Tiza Mafira, a lawyer whose work on plastic began with a 2013 petition and has since helped bring about single-use plastic bag bans in a fifth of the sprawling nation's cities and districts. Now, she wants to push for less waste in take-out meals and packaging for food and personal care products like shampoo. She knows that's harder. When people ask how it might happen, they often expect the answer to be "a better product that's more eco-friendly and as cheap and affordable as plastic," she told me. It would be simpler if that was so, but Mafira knows substitute materials tend to bring problems of their own. Indeed, while alternatives to plastic have a role to play, a wholesale switch to a different version of disposability would bring unintended consequences, like the razing of forests for wooden forks. And the words attached to some of these materials—biodegradable, compostable—are malleable, often used to manipulate us. Much compostable packaging, for example, can't degrade in a backyard heap but must be processed in an industrial-scale composter, so it makes sense only where a locality collects compostable waste. A compostable container tossed in with the recycling, or away with the trash, may do more harm than good. So, Mafira says, "what we're trying to build is the narrative that the solution is a system"—rather than a different material. A system that lets us move away from single use by making reuse possible.

Recycling has a role to play, in a limited way for plastic and more widely for easier-to-process glass, aluminum, paper, and cardboard. But it, too, is no substitute for the most fundamental answer.

At bottom, what we really need is less. Less throwaway packaging, less unnecessary junk that will quickly make its way to the rubbish bin, fewer single-use utensils and wet wipes and low-quality garments that fall apart after a handful of uses. That view is certainly where Roland Geyer—the scientist whose work has quantified all the plastic humanity has ever produced—has landed. To him, it's clear the relentless growth he's charted is "just a recipe for disaster." Every year, he told me, he updates his estimate of the cumulative total of plastic that's been made. Watching it climb, he can see plainly that the problem is "the 'how much.' For me, it's like blatantly obvious at this point." On the other side of the world, Tiza Mafira feels the same. What she hopes to see—what she's fighting to bring about—is a moment when "there is less oil production because there is less plastic production" because "there is less plastic being provided to the consumer."

It's important to remember less doesn't mean none. Industry likes to equate cutting back on throwaway cups with forgoing plastic's more meaningful uses. We mustn't fall for it. While there's truth to companies' oft-repeated reminder that some plastic is put to valuable, even essential, use, we shouldn't let it obscure the reality that a great deal of

EPILOGUE

what they foist on us is unnecessary—things we never asked for and wouldn't miss if they were gone. Like the ice cream cone drip catcher that stopped me in my tracks a few summers ago, these are items whose very existence is driven not by anyone's desire for them but solely by corporations' hunger for profit.

What's more, there are ways to ameliorate the impact of the plastic we do use. Making it doesn't have to mean vast pellet spills and toxic emissions that sicken plants' neighbors; tougher regulation can force companies to do things better. It can also push them to remove harmful chemicals from their products, and shrink plants' climate harms by running them on electricity rather than fossil fuels.

A less plastic world can seem like a utopian dream, unrealistic and out of reach. It's certainly hard to envision after decades in which the trajectory has gone only one way, and we've felt so powerless to change it. That's why I've tried in this book to show how our view of what's driving plastic's growth has been blinkered and distorted by powerful interests, working to obscure their own role. When we recognize that truth, we can finally understand the obstacle all along has not been us, but the voracious industry pushing always for more. Shifting entrenched systems is not easy, and the fossil fuel and petrochemical companies profiting from this one will not change without a fight. But with my eyes open now, I can see not only the difficulty of fixing our plastic mess but the possibility that exists for doing so. Hope, I believe, comes from honesty, and from action. Practical steps, one at a time, in a new, better direction. The changes that could get us to a less plastic future—as simple as installing more water fountains, and as complex as reimagining our food supply chains—are within our grasp.

Plastic has a special power. With its unique visibility, and its constant presence in our lives, it's one of the most tangible markers of the harm wrought by an economic model that elevates profits above all else. That makes it a guidepost we can hold to as we seek a path toward a different kind of world. One in which we finally start putting the well-being of humans and the planet that sustains us over the greedy demands of huge corporations. Because, having dug so deep into the decisions that have brought us where we are, what I am left with is an understanding that it doesn't have to be this way. Once we see clearly what this industry has done, we can finally begin to reverse it.

ACKNOWLEDGMENTS

My first thanks are owed to the many people who shared their time, expertise, and observations with me—in person, on calls, and by email—as I reported this book. Some of them are quoted in its pages, others are not, but all helped shape my understanding of the many strands of the plastic and petrochemical industry's story.

The Pulitzer Center on Crisis Reporting supported my work in Indonesia, and the McGraw Fellowship for Business Journalism, at the City University of New York's Craig Newmark Graduate School of Journalism, funded my travel to the Ohio River Valley, Texas, and Louisiana. I am immensely grateful to both, and in particular to Tom Hundley at the Pulitzer Center and Jane Sasseen at the McGraw Center for their wise counsel, and endless patience with visa delays, pandemic travel delays, and article delays. I hope they feel the end result was worth the wait. Thanks, too, to the editors who published stories drawn from my plastics reporting, and whose smart questions and comments made it—and by extension this book—stronger: Rob Kunzig at *National Geographic*; Roger

ACKNOWLEDGMENTS

Cohn, Jeremy Deaton, Fen Montaigne, and Elizabeth Royte at *Yale Environment 360*; and Andrea Thompson and Seth Fletcher at *Scientific American*.

Fadiyah Alaidrus provided invaluable help with reporting and translation, and excellent company, on my Indonesian travels. I would never have gotten a visa without the savvy and sponsorship of Kresna Astraatmadja, at Pikser Indonesia Productions, and the kind support of Arisman, at the Center for Southeast Asian Studies, in Jakarta. Thank you to Aditya Warhana for introducing me to Arisman. Daru Setyorini, Prigi Arisandi, and their team at Ecoton generously shared their time and knowledge, helping me to see plastic's impact on East Java up close. Yuyun Ismawati Drwiega provided advice and information as I planned my trip, and her team at the Nexus3 Foundation, particularly Tika, Annisa, and Nindhita, assisted with my West Java reporting—and always knew the best places to stop and eat. What a joy it was to reconnect with Eliz Winarko and Soediono in Jakarta. On another continent, huge thanks to Mila Merzagora for her skillful translation from Italian.

I couldn't possibly be more grateful to my extraordinary agent, Jessica Papin, who has worked so hard on behalf of my writing, which her wise editorial eye invariably improves. I am so lucky to have her in my corner, and the whirlwind tour of Abu Dhabi was exhilarating and unforgettable. My thanks to all her colleagues at Dystel, Goderich & Bourret too. It's been a true pleasure to work with Jacob Surpin—I'm deeply grateful to him for believing in this book, and for providing insight and guidance that helped shape it. Thanks, too, to all the rest of the team at Avery and Tarcher—Tracy Behar, Lucia Watson, Marian Lizzi, Megan Newman, and Lota Erinne. And to copy editor M. P. Klier and production editor Sally Knapp for their careful attention to the manuscript and for saving me from mistakes small and large. In the UK, I count myself lucky to be published by Octopus and Monoray; I'm so glad Jake Lingwood saw something worthwhile in this project, and am thankful to him

ACKNOWLEDGMENTS

and Jess Minocha for input that helped strengthen the manuscript, and to Pauline Bache, Megan Brown, Chloë Johnson-Hill, and Charlotte Sanders for helping it out into the world. Thanks also to Warner Bros. Discovery and HBO Max for permission to use the *Veep* quote in my epigraph, and to Tiffany Nguyen for making the process of requesting it so easy.

Carla Power helped me see there was a book in the germ of an idea about the plastic industry, and then read almost every word of it more than once, offering new wisdom every time. And I'm so grateful for the rest of my London writing crew too—Moni Mohsin, Anna Minton, Farrah Jarral, Selina Mills, and Natasha Randall—for endless support, advice, and encouragement, and for always being ready to read pages and help me see how to make them better. Being connected to all of you is a tonic for the loneliness of book writing. My dear friend Beth Harpaz gave the entire manuscript a careful and thorough read within days of my asking. Thank you so much, Beth, for both big-picture and small-bore suggestions that improved this book at the very last minute. My friendship with Amy Golden goes back to our long-ago days in Indonesia together, and even though she didn't join me on the trip I made there for this book, I could hear her wry observations in my head the whole time. Then she provided smart suggestions on the chapter set there. I can't wait for our trip back together.

I'm so lucky to have my college besties, Hannah Cooper, Kate Brownlee, and Terri Cloutier, beside me as I walk through life, and to be part of a conversation with them—in person and by text, email, and innumerable Zooms—that's been underway for more years than we might care to count. I'm so thankful too for my incredible sister, Jill Noonan, who is not only a treasured friend but also an editor extraordinaire who brought her sharp eye to *Plastic Inc.* on a tight deadline.

I have dedicated this book, with so much love, to my parents, Ronnie and Barry Gardiner. I am blessed to be your daughter, and would not be the person I am today without the love and support you have provided

ACKNOWLEDGMENTS

my entire life. I am endlessly grateful for you both. I am so proud of my daughter, Anna: Watching you grow into the amazing young woman you are has been my greatest privilege and joy. And thank you, always, to my husband and best friend, Dan Waldram. I am thankful beyond words to be sharing my life with you.

NOTES

Epigraph

vii "The hierarchy of substances . . . replace them all": Roland Barthes, *Mythologies: The Complete Edition, in a New Translation*, trans. Richard Howard and Annette Lavers (Hill and Wang, 2012), 195.

vii "Oh great, we've upset the plastics industry? This whole building is bankrolled by plastics": *Veep*, season 1, episode 1, "Fundraiser," created by Armando Iannucci, performance by Julia Louis-Dreyfus, first aired April 22, 2012, on HBO, Warner Bros. Discovery; streaming on HBO Max.

Prologue

1 The headline felt: Matthew Taylor, "$180bn Investment in Plastic Factories Feeds Global Packaging Binge," *The Guardian*, December 26, 2017, https://www.theguardian.com/environment/2017/dec/26/180bn-investment-in-plastic-factories-feeds-global-packaging-binge.

3 worthy of Broadway: Peter Mancusi, councilman in Yonkers, New York, as quoted in Linda Greenhouse, "Yonkers Studies a No-Return Ban," *New York Times*, September 9, 1971, https://www.nytimes.com/1971/09/09/archives/yonkers-studies-a-noreturn-ban-bill-would-require-refund-on-bottles.html.

4 "If plastic were a country . . . greenhouse gas emitter": Jim Vallette et al., *The New Coal: Plastics and Climate Change* (Beyond Plastics at Bennington College, October 2021), https://static1.squarespace.com/static/5eda91260bbb7e7a4bf528d8/t/616ef29221985319611a64e0/1634661022294/REPORT_The_New-Coal_Plastics_and_Climate-Change_10-21-2021.pdf.

6 more than $760 billion . . . 4 percent a year: Nikhil Kaitwade, "Plastic Market Analysis," Future Market Insights, April 25, 2025, https://www.futuremarketinsights.com/reports/plastic-market.

NOTES

Chapter 1: Plastic Dreams

11 **Three million metric tons:** All production figures cited at the start of chapters include plastic resins, fibers, and additives and were provided to the author by Roland Geyer, professor at the Bren School of Environmental Science and Management, University of California, Santa Barbara.

11 **"Waxy solid found" . . . almost illegible script:** Luigi Trossarelli and Valentina Brunella, "Polyethylene: Discovery and Growth," Università di Torino, January 2003, https://www.researchgate.net/publication/228813221_Polyethylene_discovery_and_growth.

11 **"It was a fluke":** Frank Bebbington, as quoted in Anna Jagger, "Polyethylene: Discovered by Accident 75 Years Ago," ICIS Independent Commodity Intelligence Services, May 8, 2008, https://www.icis.com/explore/resources/news/2008/05/12/9122447/polyethylene-discovered-by-accident-75-years-ago/.

12 **When heated, it could . . . polyethylene a year:** Carol Kennedy, *ICI: The Company That Changed Our Lives* (Hutchinson, 1986), 68–71.

12 **making an "almost insoluble" problem "comfortably manageable":** Robert Watson-Watt, as quoted in Claudia Flavell-While, "Dermot Manning and Colleagues at ICI—Plastic Fantastic," *Chemical Engineer*, November 1, 2011, https://www.thechemicalengineer.com/features/cewctw-dermot-manning-and-colleagues-at-ici-plastic-fantastic/.

12 **Royal Air Force pilots in 1941 . . . less portable:** Kennedy, *ICI*, 74–75.

13 **Tusks good enough . . . array of colors and textures:** Stephen Fenichell, *Plastic: The Making of a Synthetic Century* (HarperCollins, 1996), 40–47.

14 **"no longer be necessary" . . . "to mold or shape":** Susan Freinkel, *Plastic: A Toxic Love Story* (Houghton Mifflin Harcourt, 2011), 14–17.

15 **"By the way, what" . . . "create new wants":** Kennedy, *ICI*, 58–60.

15 **two potential recipes . . . based on benzene:** Fenichell, *Plastic*, 172.

16 **According to legend . . . out of the ground:** Freinkel, *Plastic*, 59.

16 **Fawcett had worked for a time at the American Petroleum Institute:** Kennedy, *ICI*, 62.

17 **"selling the manufacturer" . . . "never satisfied":** Jeffrey L. Meikle, *American Plastic: A Cultural History* (Rutgers University Press, 1995), 175–76.

18 **"could be persuaded to buy" . . . "skills of consumption":** Frederick Allen, *Only Yesterday: An Informal History of the 1920s*; and Edward Cowdrick, as quoted in Kerryn Higgs, "A Brief History of Consumer Culture," *MIT Press Reader*, January 11, 2021, https://thereader.mitpress.mit.edu/a-brief-history-of-consumer-culture/.

18 **"The future of business" . . . by 1956:** Heather Rogers, *Gone Tomorrow: The Hidden Life of Garbage* (New Press, 2005), 114, 120–24, 257.

18 **"want creation":** As quoted in Higgs, "A Brief History."

19 **"To assure itself" . . . "make for purchaser demand":** Edward Bernays, *Propaganda* (Ig Publishing, 2004), https://books.google.co.uk/books?id=3De8nd_B_C8C&pg=PA9&source=gbs_toc_r&cad=2#v=onepage&q=single%20factory&f=false, 84, 37, 74–77.

19 **For Lucky Strike . . . unhygienic:** Richard Gunderman, "The Manipulation of the American Mind: Edward Bernays and the Birth of Public Relations," *The Conversation*, July 9, 2015, https://theconversation.com/the-manipulation-of-the-american-mind-edward-bernays-and-the-birth-of-public-relations-44393.

NOTES

19 When *House Beautiful* ... juiced production further: Meikle, *American Plastic*, 172–77; and Freinkel, *Plastic*, 26–27, 57.

20 General Electric had assigned ... fastest-selling toy in US history: Fenichell, *Plastic*, 260–61.

20 The company was developing ... moved on to Frisbees: B. A. Wells and K. L. Wells, "Wham-O and Petroleum Product Hoopla," American Oil & Gas Historical Society, February 3, 2011, https://aoghs.org/products/petroleum-product-hoopla/.

21 "as to encourage" ... "without a second thought": Meikle, *American Plastic*, 176.

21 Foster Grant ... "completely discardable": Rogers, *Gone Tomorrow*, 113, 116, 121–22.

22 Ordering decades-old reports ... academic libraries: Beth Gardiner, "How an Early Oil Industry Study Became Key in Climate Lawsuits," *Yale Environment 360*, November 30, 2022, https://e360.yale.edu/features/climate-lawsuits-oil-industry-research.

24 documents showing oil companies such as ... with tobacco industry funds: "Smoke & Fumes: Smoke," Center for International Environmental Law, 2016, https://www.smokeandfumes.org/smoke.

24 "informed me that Shell has several cigaret [sic] filters and wondered concerning our possible interest": P. A. Eichorn, October 27, 1967, Philip Morris Records, Master Settlement Agreement, https://www.industrydocuments.ucsf.edu/docs/nhlf0119.

24 a researcher at tobacco firm Liggett & Myers ... "other tobacco companies": Max Samfield, "Polypropylene Fibers Produced by Enjoy Chemical Company (Esso Subsidiary)," April 3, 1968, Liggett & Myers Records, https://www.industrydocuments.ucsf.edu/docs/sgkw0009.

25 as handwritten notes ... found unfamiliar: "Exxon-Filter Meeting," July 9, 1991, R. J. Reynolds Records, Master Settlement Agreement, https://www.industrydocuments.ucsf.edu/docs/lznv0087.

25 An estimated 4.5 trillion butts: "Tobacco: Poisoning Our Planet," World Health Organization, 2022, https://apps.who.int/iris/bitstream/handle/10665/354579/9789240051287-eng.pdf?sequence=1.

25 world's most common form of plastic pollution: "Our Planet Is Choking on Plastic," United Nations Environment Programme, accessed January 18, 2024, https://www.unep.org/interactives/beat-plastic-pollution/.

25 "deadliest fraud in the history of human civilization": Robert Proctor, Stanford University professor of the history of science, as quoted in Pagan Kennedy, "Who Made That Cigarette Filter?," *New York Times*, July 6, 2012, https://www.nytimes.com/2012/07/08/magazine/who-made-that-cigarette-filter.html.

25 A confidential 1966 Philip Morris memo ... "effective advertising gimmick": W. L. Dunn and M. E. Johnston, "Market Potential of a Health Cigarette," June 1966, Philip Morris Records, Master Settlement Agreement, https://www.industrydocuments.ucsf.edu/docs/jfvc0123.

26 Hill & Knowlton ... "denial and deception": Center for International Environmental Law, "Smoke & Fumes: Smoke."

26 "worked first for oil" ... "but with the oil industry": "New Documents Reveal Denial Playbook Originated with Big Oil, Not Big Tobacco," Center for International Environmental Law, June 20, 2016, https://www.ciel.org/news/oil-tobacco-denial-playbook/#:~:text=Exxon%20and%20Shell%20patented%20and,to%20bring%20them%20to%20market.

NOTES

27 **nearly 70 percent of textiles . . . obviously artificial:** "Synthetics Anonymous: Fashion Brands' Addiction to Fossil Fuels," Changing Markets Foundation, June 2021, https://changingmarkets.org/report/synthetics-anonymous-fashion-brands-addiction-to-fossil-fuels/.

27 **half of all plastic is for single-use items . . . *Five trillion* plastic bags are used each year:** United Nations Environment Programme, "Our Planet Is Choking on Plastic."

27 **average American goes through . . . level of wastefulness:** Dominic Charles, Laurent Kimman, and Nakul Saran, *The Plastic Waste Makers Index* (Minderoo Foundation, 2021), 43, https://cdn.minderoo.org/content/uploads/2021/05/27094234/20211105-Plastic-Waste-Makers-Index.pdf.

27 **market for take-out food containers . . . growing by 3.5 percent a year:** Ismail Sutaria, "Takeaway Containers Market," Future Market Insights, April 18, 2025, https://www.futuremarketinsights.com/reports/takeaway-containers-market.

30 **"fill a hollow tooth":** ICI's Peter Allen, as quoted in Kennedy, *ICI*, 70.

30 **Other experts' projections:** World Economic Forum, Ellen MacArthur Foundation, and McKinsey & Company, *The New Plastics Economy: Rethinking the Future of Plastics* (Ellen MacArthur Foundation, 2016), https://www.ellenmacarthurfoundation.org/the-new-plastics-economy-rethinking-the-future-of-plastics.

30 **plastic production could triple by 2050:** IRENA, "Reaching Zero with Renewables," International Renewable Energy Agency, 2020, 72, https://www.irena.org/-/media/Files/IRENA/Agency/Publication/2020/Sep/IRENA_Reaching_zero_2020.pdf.

31 **world's largest producer of single-use plastics:** Dominic Charles and Laurent Kimman, *Plastic Waste Makers Index 2023* (Minderoo Foundation, 2023), 27, https://cdn.minderoo.org/content/uploads/2023/02/04205527/Plastic-Waste-Makers-Index-2023.pdf.

31 **acknowledges electric . . . by 2050:** "Energy Demand Trends," ExxonMobil, August 26, 2024, https://corporate.exxonmobil.com/sustainability-and-reports/global-outlook/energy-demand-trends.

31 **company said it expected chemical production to increase by 80 percent over the period:** "Global Outlook, Energy Supply," ExxonMobil, January 8, 2024, https://web.archive.org/web/20240129164639/https://corporate.exxonmobil.com/what-we-do/energy-supply/global-outlook/energy-supply#Liquids.

31 **pump even more oil in 2050 than it does today:** "ExxonMobil Global Outlook: Executive Summary; Our View to 2050," ExxonMobil, accessed January 19, 2024, https://corporate.exxonmobil.com/-/media/global/files/global-outlook/2023/2023-global-outlook-executive-summary.pdf.

31 **"Chemicals and commercial transportation account for almost all of the [oil] demand growth":** "Global Outlook, Energy Supply," ExxonMobil.

31 **"as more of the world's . . . essential for modern living":** "Global Outlook, Energy Demand: Three Drivers," ExxonMobil, January 8, 2024, https://web.archive.org/web/20240111095237/https://corporate.exxonmobil.com/what-we-do/energy-supply/global-outlook/energy-demand#Industrial.

31 **while use of oil . . . going forward:** Yuya Akizuki et al., *Oil 2023: Analysis and Forecast to 2028* (International Energy Agency, June 2023), https://iea.blob.core.windows.net/assets/6ff5beb7-a9f9-489f-9d71-fd221b88c66e/Oil2023.pdf.

32 **80 percent of petrochemicals:** Eren Çetinkaya et al., "Petrochemicals 2030: Reinventing the Way to Win in a Changing Industry," McKinsey & Company, February 21, 2018,

NOTES

https://www.mckinsey.com/industries/chemicals/our-insights/petrochemicals-2030-reinventing-the-way-to-win-in-a-changing-industry.

32 **Shell in 2021 . . . "want and need":** "Shell Accelerates Drive for Net-Zero Emissions with Customer-First Strategy," Shell Global, February 11, 2021, https://www.shell.com/media/news-and-media-releases/2021/shell-accelerates-drive-for-net-zero-emissions-with-customer-first-strategy.html; Charles, Kimman, and Saran, *Plastic Waste Makers Index*, 41; and "Our Growth Projects," Shell Global, accessed January 19, 2024, https://www.shell.com/business-customers/chemicals/about-shell-chemicals/our-growth-projects.html.

32 **By 2025 . . . "adequate returns":** Alex Scott, "Shell to Pull Back from Chemicals," *Chemical and Engineering News*, March 26, 2025, https://cen.acs.org/business/petrochemicals/Shell-pull-back-chemicals/103/web/2025/03.

32 **largest single driver of oil-demand growth this decade:** Araceli Fernandez Pales et al., *The Future of Petrochemicals* (International Energy Agency, October 2018), https://www.iea.org/reports/the-future-of-petrochemicals.

Chapter 2: "An Excellent Salesman"

35 **Adolphus Green never meant . . . in charge:** Gary Hoover, "Uneeda Business History: The Nabisco Story," American Business History Center, September 12, 2021, https://americanbusinesshistory.org/uneeda-business-history-the-nabisco-story/.

35 **he decided the company . . . butter and soap:** Diana Twede, "Uneeda Biscuit: The First Consumer Package?," *Journal of Macromarketing* 17, no. 2 (1997): 82–88, https://doi.org/10.1177/027614679701700208.

37 **"motivating sales" . . . "product lines and profitability":** Eric Outwater, of Stuart and Gunn, in *Proceedings: First National Conference on Packaging Wastes* (Solid Waste Management Office, US Environmental Protection Agency, 1971), 3–5, https://nepis.epa.gov/Exe/ZyNET.exe/2000Q54D.TXT?ZyActionD=ZyDocument&Client=EPA&Index=Prior+to+1976&Docs=&Query=&Time=&EndTime=&SearchMethod=1&TocRestrict=n&Toc=&TocEntry=&QField=&QFieldYear=&QFieldMonth=&QFieldDay=&IntQFieldOp=0&ExtQFieldOp=0&XmlQuery=&File=D%3A%5Czyfiles%5CIndex%20Data%5C70thru75%5CTxt%5C00000001%5C2000Q54D.txt&User=ANONYMOUS&Password=anonymous&SortMethod=h%7C-&MaximumDocuments=1&FuzzyDegree=0&ImageQuality=r75g8/r75g8/x150y150g16/i425&Display=hpfr&DefSeekPage=x&SearchBack=ZyActionL&Back=ZyActionS&BackDesc=Results%20page&MaximumPages=1&ZyEntry=1&SeekPage=x&ZyPURL.

37 **Head office was happy . . . just sixteen years:** Bartow J. Elmore, *Citizen Coke: The Making of Coca-Cola Capitalism* (W. W. Norton, 2015), 225–28, 231; Catherine Lerza, "Administration 'Pitches In' to Outlaw Throwaways," *Environmental Action*, May 25, 1974, 5, http://michiganintheworld.history.lsa.umich.edu/environmentalism/files/original/52eaa3b77dd75561d53a31bb7edea32a.pdf; and John Stuart, "C&C Super Corp. to Open Third Plant Next Month to Can Soft Drinks in Chicago," *New York Times*, April 25, 1954, F1, https://timesmachine.nytimes.com/timesmachine/1954/04/25/issue.html.

39 **"They're a pain in the neck":** David Bird, "Returnable Bottles Are Proving Unpopular with Dealers Here," *New York Times*, February 2, 1971, https://www.nytimes.com/1971/02/02/archives/returnable-bottles-are-proving-unpopular-with-dealers-here.html.

NOTES

39 "recovery burden of sixty-five thousand truckloads": Chokola, as quoted in Elmore, *Citizen Coke*, 231.

39 "The consumer is the one who pays the taxes to clean up the mess": Chokola, as quoted in David Bird, "Bottler Fights a Trend, Pleads for Returnables," *New York Times*, January 7, 1972, https://www.nytimes.com/1972/01/07/archives/bottler-fights-a-trend-pleads-for-returnables-bottler-leads-fight.html.

39 Coke was 30 to 40 percent . . . Lucian Smith; "offer the" . . . he acknowledged: "Hearings Before the Subcommittee on Antitrust and Monopoly of the Committee on the Judiciary, United States Senate, Ninety-Second Congress, Second Session," August 8–10 and September 12 and 14, 1972, in *Hearings, Reports and Prints of the Senate Committee on the Judiciary* (US Government Printing Office, 1971), 164, https://books.google.co.uk/books?id=gBA2AAAAIAAJ&printsec=frontcover#v=onepage&q&f=false.

39 In Richmond, Virginia . . . fifty-nine cents: "Your Can Is a Waste: A Report on the Problems Caused by the Introduction of Non-Returnable Containers of Beer and Soda," testimony by Barry L. Goldstein, assistant director of legislative affairs, College Young Democrats of America, *Solid Waste Management Act of 1972: Hearings Before the Subcommittee on the Environment of the Committee on Commerce, United States Senate, 92nd Congress* (US Government Printing Office, 1972), 488–501, https://books.google.co.uk/books?id=5BcQAAAAIAAJ&printsec=frontcover#v=onepage&q&f=false.

40 "have removed" the cheaper option "from the realm of consumer choice": Lerza, "Administration 'Pitches In,'" 5.

40 137 billion plastic bottles: Coca-Cola Company, 2023 Environmental Report, 9, https://www.coca-colacompany.com/content/dam/company/us/en/reports/2023-environmental-update/2023-environmental-update.pdf.

40 Chemical giant DuPont boasted . . . raking in $20 million: Jeffrey L. Meikle, "Material Doubts: The Consequences of Plastic," *Environmental History* 2, no. 3 (1997): 278–300, https://doi.org/10.2307/3985351.

40 The Standard Packaging company tripled . . . take-out containers, and more: Heather Rogers, *Gone Tomorrow: The Hidden Life of Garbage* (New Press, 2005), 117, 122.

41 "The future of plastics" . . . "*billions* of units": Lloyd Stouffer, "Plastics Packaging: Today and Tomorrow," remarks prepared for delivery at 1963 National Plastics Conference, Sheraton-Chicago Hotel, Chicago, Society of the Plastics Industry, accessed September 21, 2024, https://discardstudies.com/wp-content/uploads/2014/07/stoffer-plastics-packacing-today-and-tomorrow-1963.pdf. Italics in original.

41 "teach customers how to waste": Rebecca Altman, "American Beauties," *Topic*, August 2018, https://static1.squarespace.com/static/5703f76762cd94e407457a23/t/6509a958a269415445c55d2e/1695131997312/Am+Beauties+Topic+2018+copy.pdf.

41 Three years after the first squeeze bottle . . . shampoo: Meikle, *American Plastic*, 190.

41 "You are filling" . . . "tripling by 1970": Stouffer, "Plastics Packaging."

42 "Major U.S. Cities" . . . $2 million a day for five years: Gladwin Hill, "Major U.S. Cities Face Emergency in Trash Disposal," *New York Times*, June 16, 1969, https://www.nytimes.com/1969/06/16/archives/major-us-cities-face-emergency-in-trash-disposal-growing-national.html.

42 equivalent of ninety-two thousand barrels of oil . . . Central American nations: Lerza, "Administration 'Pitches In,'" 4.

NOTES

42 can and bottle producers . . . "plastic-coated materials": *Proceedings: First National Conference on Packaging Wastes* (Solid Waste Management Office, US Environmental Protection Agency, 1971), 1–10, 14–15, 85–86, 231–42.

43 He'd been just . . . president of the company: Leonard Stefanelli, "Everything You Wanted to Know About 'Garbage' but Were Afraid to Ask," San Francisco Historical Society, reprinted from *The Argonaut* 25, no. 1 (Summer 2014), https://www.sfhistory.org/everything-you-wanted-to-know-about-garbage-but-were-afraid-to-ask/.

44 He described how . . . "unlikely to happen": *Proceedings: First National Conference on Packaging Wastes*, 37–42, 121, 14–15.

46 clients had included Coca-Cola: Arsen Darnay and Gary Nuss, "Environmental Impacts of Coca-Cola Beverage Containers," appendix C, exhibit 1, of *Technology Assessment Activities in the Industrial, Academic and Governmental Communities*, Hearings Before the Technology Assessment Board of the Office of Technology Assessment, Congress of the United States, Ninety-Fourth Congress, Second Session, December 1976, https://repository.digital.georgetown.edu/handle/10822/708594?q=Roosevelt%20Civil%20War%20Envelopes%20Collection%20;%20page%205,%20image%203.

46 suggested that a fee . . . "the swing can be too far": *Proceedings: First National Conference on Packaging Wastes*, 49–51, 107–13.

47 In Atlanta . . . empty cans and bottles: Finis Dunaway, *Seeing Green: The Use and Abuse of American Environmental Images* (University of Chicago Press, 2015), 86; and Elmore, *Citizen Coke*, 238.

47 "We often discard today" . . . "public subsidy of waste": Richard Nixon, "Special Message to the Congress on Environmental Quality," American Presidency Project, February 10, 1970, https://www.presidency.ucsb.edu/documents/special-message-the-congress-environmental-quality.

47 "really end the industry": Sidney Gross, as quoted in Meikle, "Material Doubts."

48 In billboards . . . "good outdoor manners": Rogers, *Gone Tomorrow*, 142.

48 "how-to-do-it" litter-reduction kits . . . empties out windows: "Progress Is Noted in U.S. Cleanup," *New York Times*, February 24, 1959, https://www.nytimes.com/1959/02/24/archives/progress-is-noted-in-u-s-cleanup-keep-america-beautiful-inc-reports.html; and "Industry Fosters Drive Against Litter; $400,000 Voted to Better U.S. Habits," *New York Times*, October 14, 1954, https://www.nytimes.com/1954/10/14/archives/industry-fosters-drive-against-litter-400000-voted-to-better-u-s.html.

48 "Keeping America clean and beautiful is your job": Elmore, *Citizen Coke*, 234.

48 "alibi" . . . "stop shifting the blame": Dunaway, *Seeing Green*, 93.

48 "packages don't litter, people do": Rogers, *Gone Tomorrow*, 144.

48 "Bend a Little" . . . "to go along": Elmore, *Citizen Coke*, 239; and "If you love me, don't leave me," Coca-Cola ad, *New York Times*, 33, https://timesmachine.nytimes.com/timesmachine/1970/04/22/issue.html.

49 "Our 'soft sell'" . . . slam litterers as "slobs": Ginger Strand, "The Crying Indian," *Orion*, November/December 2008, https://orionmagazine.org/article/the-crying-indian/.

49 "People who had been working" . . . "propagandistic": Dunaway, *Seeing Green*, 79–95.

50 "co-opted the icon of resistance" . . . "the ideology of waste": Strand, "The Crying Indian."

NOTES

51 "laws will not be enacted" ... "result of organic human development": Rogers, *Gone Tomorrow*, 144, 27.

52 A student committee appointed ... for a vote: Greenhouse, "Yonkers Studies a No-Return Ban"; and Goldstein, "Your Can Is a Waste."

53 polled at 80 percent ... Edgewater, New Jersey: Goldstein, "Your Can Is a Waste"; and Taylor H. Bingham and Paul F. Mulligan, *The Beverage Container Problem: Analysis and Recommendations*, prepared for the Office of Research and Monitoring, US Environmental Protection Agency (US Government Printing Office, September 1972), 1, https://nepis.epa.gov/Exe/ZyNET.exe/9101ESAG.txt?ZyActionD=ZyDocument&Client=EPA&Index=Prior%20to%201976&Docs=&Query=&Time=&EndTime=&SearchMethod=1&TocRestrict=n&Toc=&TocEntry=&QField=&QFieldYear=&QFieldMonth=&QFieldDay=&UseQField=&IntQFieldOp=0&ExtQFieldOp=0&XmlQuery=&File=D%3A%5CZYFILES%5CINDEX%20DATA%5C70THRU75%5CTXT%5C00000021%5C9101ESAG.txt&User=ANONYMOUS&Password=anonymous&SortMethod=h%7C-&MaximumDocuments=1&FuzzyDegree=0&ImageQuality=r75g8/r75g8/x150y150g16/i425&Display=hpfr&DefSeekPage=x&SearchBack=ZyActionL&Back=ZyActionS&BackDesc=Results%20page&MaximumPages=1&ZyEntry=2.

53 slammed bottle-bill advocates ... 91 percent of Oregonians approved: Rogers, *Gone Tomorrow*, 147, 150.

54 The same day as ... the job-killer: Ronald Sullivan, "Jersey Plan May Limit Containers," *New York Times*, September 9, 1971, https://www.nytimes.com/1971/09/09/archives/jersey-plan-may-limit-containers-hearing-ordered-in-new-jersey-on.html; and "Jersey Delays Bill for Ban on No-Return Containers," *New York Times*, September 23, 1971, https://www.nytimes.com/1971/09/23/archives/jersey-delays-bill-for-ban-on-noreturn-containers-jersey-panel.html.

54 save the equivalent ... "by their free choice of containers": Rogers, *Gone Tomorrow*, 151.

54 "Left unsaid ... the company's waste": Elmore, *Citizen Coke*, 255.

55 "Bottle Deposit Proposal Fizzles Out in Legislature": Dyer Oxley, "Washington Bottle Deposit Proposal Fizzles Out in Legislature," KUOW, February 14, 2024, https://www.kuow.org/stories/washington-bottle-deposit-proposal-fizzles-out-in-legislature#.

55 "many bottles have NO DEPOSIT ... on the bottom": Rogers, *Gone Tomorrow*, 148.

55 Coke argued the scanning technology ... how much they owed: Finn Arne Jørgensen, *Making a Green Machine*, as quoted in Elmore, *Citizen Coke*, 255–56.

55 "returning your empty bottles" ... heading to the grocery store: Finn Arne Jørgensen, "A Pocket History of Bottle Recycling," *The Atlantic*, February 27, 2013, https://www.theatlantic.com/technology/archive/2013/02/a-pocket-history-of-bottle-recycling/273575/.

55 gets 98 percent back: Irene Banos Ruiz and Jeannette Cwienk, "A Look at Germany's Bottle Deposit Scheme," *Deutsche Welle*, November 17, 2021, https://www.dw.com/en/how-does-germanys-bottle-deposit-scheme-work/a-50923039.

55 New South Wales lawmaker ... dropped their bottle bill: Kirsty Needham, "Millions of Dollars in Attack Ads Coming in March," *Sydney Morning Herald*, October 24, 2014, https://www.smh.com.au/national/nsw/millions-of-dollars-in-attack-ads-coming-in-march-20141024-11b5pi.html.

NOTES

56   In Edinburgh . . . three-quarters of Scots supported the idea: Maeve McClenaghan, "Investigation: Coca Cola and the 'Fight Back' Against Plans to Tackle Plastic Waste," *Unearthed*, https://unearthed.greenpeace.org/2017/01/25/investigation-coca-cola-fight-back-plans-tackle-plastic-waste/.

56   "shut down debate" . . . "No one's talking about that now": Strand, "The Crying Indian."

Chapter 3: America's Plastic Boom

58   more than six hundred refineries and chemical plants: "Chemical Industry Overview," Greater Houston Partnership, April 26, 2021, https://web.archive.org/web/20240829012857/https://www.houston.org/houston-data/chemical-industry-overview.

59   researchers found dangerous heavy metals including arsenic, lead, and chromium: Garett Sansom et al., "Confirming the Environmental Concerns of Community Members Utilizing Participatory-Based Research in the Houston Neighborhood of Manchester," *International Journal of Environmental Research and Public Health* 13, no. 9 (August 2016): 839, https://doi.org/10.3390/ijerph13090839.

59   One analysis found Manchester . . . city average: Ronald White et al., "Double Jeopardy in Houston: Acute and Chronic Chemical Exposures Pose Disproportionate Risks for Marginalized Communities," Union of Concerned Scientists and Texas Environmental Justice Advocacy Services, accessed December 7, 2023, 13, https://www.ucsusa.org/sites/default/files/attach/2016/10/ucs-double-jeopardy-in-houston-full-report-2016.pdf.

59   more than one hundred tons of toxic vapors: Jordan Blum, "Harvey Exposes Another Gap Between Rich and Poor," *Houston Chronicle*, January 9, 2018, https://www.houstonchronicle.com/business/energy/article/Harvey-exposed-another-gap-between-rich-and-poor-12475880.php.

59   benzene to spike alarmingly in Manchester: Hiroko Tabuchi, "High Levels of Carcinogen Found in Houston Area After Harvey," *New York Times*, September 6, 2017, https://www.nytimes.com/2017/09/06/us/harvey-houston-valero-benzene.html.

59   fire broke out at a 242-tank . . . xylene, toluene, and pygas: *Storage Tank Fire at Intercontinental Terminals Company, LLC (ITC) Terminal: Investigation Report* (US Chemical Safety and Hazard Investigation Board, July 6, 2023), https://www.csb.gov/assets/1/6/itc_report_-_final_(july_6,_2023).pdf.

60   "catastrophic, unlike anything we've seen in modern history in this country": Harris County Attorney Christian Menefee, interview with author, January 9, 2023, as quoted in Beth Gardiner, "Deep in the Heart of Texas, an Uphill Fight for Clean Air for All," *Yale Environment 360*, April 11, 2023, https://e360.yale.edu/features/christian-menefee-houston-petrochemicals-interview.

64   In one case . . . "make a decision": Hiroko Tabuchi, "E.P.A. Offers a Way to Avoid Clean-Air Rules: Send an Email," *New York Times,* March 27, 2025, https://www.nytimes.com/2025/03/27/climate/epa-air-pollution-exemption-mercury-coal-ash.html.

64   "some level of risk" . . . acceptable levels: Richard Richter, spokesman for Texas Commission on Environmental Quality, email to author, November 3, 2023.

64   regards anything less than . . . dangerous: Katie Watkins, "A Petrochemical Company Wants to Expand in Houston's East End. The City Says Air Pollution Levels Are Already Too High," *Houston Public Media*, April 13, 2022, https://www.houstonpublicmedia.org

NOTES

/articles/news/energy-environment/2022/04/13/422601/air-pollution-levels-already-too-high-near-proposed-east-end-plant-expansion-says-houston-health-department/.

67 **"It's absolutely extraordinary" . . . "predicted this"**: Neil Chapman, ExxonMobil senior vice president, as quoted in Jordan Blum, "How the Ethane Molecule, Found in Texas Shale Fields, Changed the Gulf Coast—and the World," *Houston Chronicle*, September 18, 2018, https://www.houstonchronicle.com/business/energy/article/How-a-molecule-changed-the-Gulf-Coast-and-the-13225037.php.

67 **$10 billion plant near Corpus Christi . . . 1.7 million tons annually**: "Growing the Gulf," ExxonMobil, accessed December 7, 2023, https://corporate.exxonmobil.com/locations/united-states/growing-the-gulf.

68 **North America's virgin plastic production jumped by 60 percent**: Kathy Hipple and Anne Keller, "Updated Economics for Virgin Plastics," Ohio River Valley Institute, December 2023, https://ohiorivervalleyinstitute.org/updated-economics-for-virgin-plastics/.

68 **upward of 680 million kilograms**: Megan Quinn, "Plastic Supply Pressures Create Market Challenges and Opportunities for Recycled Resins, Analysts Say," *Waste Dive*, March 9, 2022, https://www.wastedive.com/news/plastic-supply-pressures-create-market-challenges-and-opportunities-for-rec/620081/.

68 **more than 1,300 containers-full**: "Top Containerized Export Commodities Port Houston Container Volume," Port Houston, accessed December 7, 2023, https://porthouston.com/wp-content/uploads/2023/08/4-Container-Volume-by-Commodity-2022.pdf.

68 **"Those overseas markets" . . . "connecting those dots"**: ExxonMobil, "Growing the Gulf."

69 **Japanese people . . . do better on IQ tests**: "Statement of Dr. Michael Honeycutt, chief toxicologist, Texas Commission on Environmental Quality," October 4, 2011, *Hearing Before the Subcommittee on Energy and Environment, Committee on Science, Space, and Technology*, House of Representatives, One Hundred Twelfth Congress, First Session, 27, https://www.govinfo.gov/content/pkg/CHRG-112hhrg70587/pdf/CHRG-112hhrg70587.pdf.

69 **Two-thirds of his office's . . . "in my life"**: Lisa Song and Rosalind Adams, "Texas Weakens Chemical Exposure Guidelines, Opens Door for Polluters," *Center for Public Integrity* and *Inside Climate News*, December 18, 2014, https://publicintegrity.org/environment/texas-weakens-chemical-exposure-guidelines-opens-door-for-polluters/.

69 **keynote speech at a conference**: "2018 GlobalChem Keynote, Agenda Announced," American Chemistry Council, January 31, 2021, https://www.americanchemistry.com/chemistry-in-america/news-trends/press-release/2018/2018-globalchem-keynote-agenda-announced.

69 **When he and his team . . . news site *The Intercept***: Sharon Lerner, "The War on the War on Cancer," *The Intercept*, January 12, 2020, https://theintercept.com/2020/01/12/cancer-trump-administration-epa-carcinogens-regulations/.

70 **"We're in a state" . . . "for their mistakes"**: As quoted in Gardiner, "Deep in the Heart of Texas."

70 **worst ever for cities and counties**: Christopher Hooks, "The Bonnen Tape Is Tawdry, Shocking, and Kinda Funny," *Texas Monthly*, October 15, 2019, https://www.texasmonthly.com/news-politics/dennis-bonnen-recording-tawdry-shocking/.

NOTES

71 **most heavily fracked, home to more than 1,500 wells:** Kristina Marusic, "Fractured: Harmful Chemicals and Unknowns Haunt Pennsylvanians Surrounded by Fracking," *Environmental Health News*, March 1, 2021, https://www.ehn.org/fractured-harmful-chemicals-fracking-2650428324.html.

72 **one thousand different chemicals:** Ashley L. Bolden et al., "Exploring the Endocrine Activity of Air Pollutants Associated with Unconventional Oil and Gas Extraction," *Environmental Health* 17, no. 1 (2018), https://doi.org/10.1186/s12940-018-0368-z.

72 **more than fourteen million liters:** Andrew J. Kondash, Elizabeth Albright, and Avner Vengosh, "Quantity of Flowback and Produced Waters from Unconventional Oil and Gas Exploration," *Science of the Total Environment* 574 (January 2017): 314–21, https://doi.org/10.1016/j.scitotenv.2016.09.069.

73 **funding renovations . . . charities:** Candy Woodall, "Despite Downturn, Oil and Gas Industry Still a Boost to Pa. Nonprofits," *PennLive/Patriot-News*, March 22, 2016, https://www.pennlive.com/news/2016/03/despite_downturn_oil_and_gas_i.html.

73 **one-hundred-plus chemicals known to pollute air around frack sites:** Bolden et al., "Exploring the Endocrine Activity of Air Pollutants."

73 **Researchers also . . . commonly used in fracking:** Marusic, "Fractured."

73 **Between 2008 and 2018 . . . would have been eight:** David Templeton and Don Hopey, "Are the 27 Cases of Ewing Sarcoma near Pittsburgh a Cluster?," *Pittsburgh Post-Gazette*, May 14, 2019, https://newsinteractive.post-gazette.com/ewing-sarcoma-cancer-cluster-pittsburgh-washington-westmoreland/.

73 **five to seven times . . . eight kilometers from one:** "PA Health and Environment Study: Childhood Cancer," Pennsylvania Department of Health, August 2023, accessed December 4, 2023, https://www.pa.gov/content/dam/copapwp-pagov/en/health/documents/topics/documents/environmental-health/Pitt_Summary_FNL_CANCER-%2008-15-23.pdf.

74 **Yale researchers . . . leukemia:** Cassandra J. Clark et al., "Unconventional Oil and Gas Development Exposure and Risk of Childhood Acute Lymphoblastic Leukemia: A Case–Control Study in Pennsylvania, 2009–2017," *Environmental Health Perspectives* 130, no. 8 (August 2022), https://doi.org/10.1289/ehp11092.

74 **Harvard team . . . early death:** Longxiang Li et al., "Exposure to Unconventional Oil and Gas Development and All-Cause Mortality in Medicare Beneficiaries," *Nature Energy* 7, no. 2 (January 2022): 177–85, https://doi.org/10.1038/s41560-021-00970-y.

74 **The industry, saying . . . "internet click-bait":** "Latest Yale Headline-Grabbing Study Contradicts Previous Research," Marcellus Shale Coalition, August 19, 2022, https://marcelluscoalition.org/latest-yale-headline-grabbing-study-contradicts-previous-research/.

74 **elevated rates of heart failure:** Tara P. McAlexander et al., "Unconventional Natural Gas Development and Hospitalization for Heart Failure in Pennsylvania," *Journal of the American College of Cardiology* 76, no. 24 (December 2020): 2862–74, https://doi.org/10.1016/j.jacc.2020.10.023.

74 **breathing and skin problems . . . driver of warming:** Irena Gorski and Brian S. Schwartz, "Environmental Health Concerns from Unconventional Natural Gas Development," *Oxford Research Encyclopedia of Global Public Health*, February 25, 2019, https://doi.org/10.1093/acrefore/9780190632366.013.44.

74 **one thousand new wells:** Nick Cunningham, "A Fracking-Driven Industrial Boom Renews Pollution Concerns in Pittsburgh," *Yale Environment 360*, March 21, 2019, https://

NOTES

e360.yale.edu/features/a-fracking-driven-industrial-boom-renews-pollution-concerns-in-pittsburgh.

74 **"You have to drill . . . keep drilling the wells":** Rebecca Scott, associate professor of environmental sociology, University of Missouri, as quoted in Kristina Marusic, "The Titans of Plastic," *Environmental Health News,* September 15, 2022, https://www.ehn.org/the-titans-of-plastic-2657986993.html.

75 **"There's a warmth and a caring" . . . they knew exactly what had happened:** As quoted in Beth Gardiner, "Amid Hopes and Fears, a Plastics Boom in Appalachia Is on Hold," *Yale Environment 360,* April 13, 2022, https://e360.yale.edu/features/plans-to-make-appalachia-a-plastics-hub-face-growing-hurdles.

76 **"fatigue for . . . unflattering conclusions":** Curtis Smith, head of Americas media at Shell, email to author, April 23, 2021.

77 **about 400,000 liters . . . vinyl flooring:** Hiroko Tabuchi, "Texas to New Jersey: Tracking the Toxic Chemicals in the Ohio Train Inferno," *New York Times,* April 17, 2023, https://www.nytimes.com/2023/04/17/climate/train-fire-palestine-plastics-pvc.html.

77 **"My husband and I" . . . "we are leaving":** From author's follow-up Zoom interview with Gdula, November 13, 2023.

77 **150 petrochemical plants and refineries:** "Environmental Racism in Louisiana's 'Cancer Alley' Must End, Say UN Human Rights Experts," *UN News,* United Nations, March 2, 2021, https://news.un.org/en/story/2021/03/1086172#.

78 **nation's highest concentration of millionaires:** Anya Groner, "Louisiana Chemical Plants Are Thriving off of Slavery," *The Atlantic,* May 7, 2021, https://www.theatlantic.com/culture/archive/2021/05/louisiana-chemical-plants-thriving-off-slavery/618769/.

78 **"You can get a thousand acres . . . plantation owner":** Joy Banner, cofounder and codirector of the Descendants Project, interview with author, January 5, 2023.

78 **"instead of plantations" . . . "with plants":** Groner, "Louisiana Chemical Plants."

79 **"You don't have a grocery store" . . . "just diminished":** Myrtle Felton, cofounder and co-executive director of Inclusive Louisiana, interview with author, January 5, 2023.

81 **highest risk of cancer . . . fifty times the national average:** Sharon Lerner, "When Pollution Is a Matter of Life and Death," *New York Times,* June 22, 2019, https://www.nytimes.com/2019/06/22/opinion/sunday/epa-carniogens.html.

81 **averaging up to ten times . . . by the latter half of the 2010s:** Lilian S. Dorka, deputy assistant administrator for external civil rights, US Environmental Protection Agency, Office of Environmental Justice and External Civil Rights, letter to Dr. Chuck Carr Brown, secretary, Louisiana Department of Environmental Quality, and Dr. Courtney N. Phillips, secretary, Louisiana Department of Health, October 12, 2022, 25–28, https://www.epa.gov/system/files/documents/2022-10/2022%2010%2012%20Final%20Letter%20LDEQ%20LDH%2001R-22-R6%2C%2002R-22-R6%2C%2004R-22-R6.pdf.

81 **forty-five dangerous pollutants in the parish's air:** Sharon Lerner, "A Tale of Two Toxic Cities," *The Intercept,* February 24, 2019, https://theintercept.com/2019/02/24/epa-response-air-pollution-crisis-toxic-racial-divide/.

81 **they prefer chloroprene concentrations to be . . . are Black:** Dorka, letter to Brown and Phillips, 14–15.

82 **no evidence of increased illness near its plant:** Lisa Song and Lylla Younes, "EPA Finally Calls Out Environmental Racism in Louisiana's Cancer Alley," *Grist* and *ProPublica,*

298

NOTES

October 19, 2022, https://grist.org/regulation/epa-finally-calls-out-environmental-racism-in-louisiana-cancer-alley/.

82 notes it's cut emissions by 85 percent . . . "look for ways to improve": Jim Harris, "Denka Performance Elastomer to Request EPA Act on Updated Risk Model for Chloroprene," Denka Performance Elastomer LLC, March 1, 2021, http://denka-pe.com/wp-content/uploads/2021/03/Denka-Performance-Elastomer-to-request-EPA-act-on-updated-risk-model-for-chloroprene.pdf.

82 "there is no question" . . . "elevated cancer risk": Dorka, letter to Brown and Phillips.

Chapter 4: "Guilt Eraser"

83 "Guilt Eraser": Susan Freinkel, *Plastic: A Toxic Love Story* (Houghton Mifflin Harcourt, 2011), 162.

85 Infinity . . . finite space: Finis Dunaway, *Seeing Green: The Use and Abuse of American Environmental Images* (University of Chicago Press, 2015), 99.

85 "as old as thrift" . . . "would have many lives": Oliver Franklin-Wallis, *Wasteland: The Dirty Truth About What We Throw Away, Where It Goes, and Why It Matters* (Simon & Schuster, 2023), 46–47.

86 "well-meaning but misinformed authorities": Joel Frados, "There's Something in the Air," *Modern Plastics* 44, October 1966, as quoted in Jeffrey L. Meikle, *American Plastic: A Cultural History* (Rutgers University Press, 1995), 265.

86 a handful of municipalities . . . offered the service: Heather Rogers, *Gone Tomorrow: The Hidden Life of Garbage* (New Press, 2005), 139.

86 "When it started, recycling was a pretty radical idea": Martin Bourque, executive director, Ecology Center of Berkeley, California, interview with author, July 31, 2023.

86 "the only ecologically sensible long-term solution" for a nation "knee-deep in garbage": Environmentalist Garrett De Bell, as quoted in Kate Yoder, "How the Recycling Symbol Lost Its Meaning," *Grist*, June 12, 2024, https://grist.org/culture/recycling-symbol-logo-plastic-design/.

86 In 1971, Coke's New York . . . "discarded cans": Martin Gansberg, "Recycling Drive Gains Impetus on 2d Saturday," *New York Times*, March 28, 1971, https://www.nytimes.com/1971/03/28/archives/recycling-drive-gains-impetus-on-2d-saturday.html.

86 "The plastics industry is at work" . . . collapsed into a river: The People of the State of California, ex rel. Rob Bonta, Attorney General of California, v. ExxonMobil Corporation, Complaint for Abatement, Equitable Relief, and Civil Penalties, Superior Court of the State of California, County of San Francisco, September 23, 2024, https://oag.ca.gov/system/files/attachments/press-docs/Complaint_People%20v.%20Exxon%20Mobil%20et%20al.pdf.

87 "Recycling so far" . . . was expanded: Bartow J. Elmore, *Citizen Coke: The Making of Coca-Cola Capitalism* (W. W. Norton, 2015), 246–47.

87 Taxpayers, one environmentalist . . . "institutionalizing waste generation": Patricia Taylor of Environmental Action, as quoted in Rogers, *Gone Tomorrow*, 140, 172.

88 "offer environmental advantages when disposed of in dumps and landfills": People of the State of California v. ExxonMobil.

NOTES

88 "Recycle plastic packaging?" . . . in 1971: Judd H. Alexander of the American Can Company, speaking at Packaging Institute's annual forum in 1971, as quoted in Davis Allen et al., *The Fraud of Plastic Recycling* (Center for Climate Integrity, February 2024), https://climateintegrity.org/uploads/media/Fraud-of-Plastic-Recycling-2024.pdf.

88 "The public almost" . . . "the perception issue": F. E. Krause, director environmental solutions, Geon Vinyl division, B. F. Goodrich Co., "PVC Recycling—an Overview. Presentation to: The Vinyl Industry Tripartite Meeting," September 3–4, 1992, in Merlin Chowkwanyun, Gerald Markowitz, and David Rosner, "Toxic Docs: Version 1.0," Columbia University and City University of New York, 2018, https://www.toxicdocs.org/d/91wxG1YnjQ8KjOnZ3jE9wLxg7?lightbox=1#.

88 "Recycling sounds like" . . . because it was "uneconomical": People of the State of California v. ExxonMobil.

89 **More than ten thousand chemicals**: Helene Wiesinger, Zhanyun Wang, and Stefanie Hellweg, "Deep Dive into Plastic Monomers, Additives, and Processing Aids," *Environmental Science and Technology* 55, no. 13 (June 2021): 9339–51, https://doi.org/10.1021/acs.est.1c00976.

89 "I mean, who's . . . pay to do that?": Emily Kwong and Laura Sullivan, "The Myth of Plastic Recycling," *Short Wave*, NPR, December 12, 2022, https://www.npr.org/transcripts/1141601301.

90 **leached 150 different chemicals into drinks**: Spyridoula Gerassimidou et al., "Unpacking the Complexity of the PET Drink Bottles Value Chain: A Chemicals Perspective," *Journal of Hazardous Materials* 430 (May 2022): 128410, https://doi.org/10.1016/j.jhazmat.2022.128410.

90 **reported none were free from harmful chemicals**: Sara Brosché et al., "Widespread Chemical Contamination of Recycled Plastic Pellets Globally," International Pollutants Elimination Network, December 2021, https://ipen.org/documents/widespread-chemical-contamination-recycled-plastic-pellets-globally.

90 **toys, kitchen utensils, and other common items**: A. Guzzonato, F. Puype, and S. J. Harrad, "Evidence of Bad Recycling Practices: BFRs in Children's Toys and Food-Contact Articles," *Environmental Science: Processes and Impacts* 19, no. 7 (July 2017): 956–63, https://doi.org/10.1039/c7em00160f.

90 "The aims of" . . . "avenue of approach": *Proceedings: First National Conference on Packaging Wastes* (Solid Waste Management Office, US Environmental Protection Agency, 1971), 5, 7, 15–16.

91 "and when you get . . . a lot less money": Mobil Oil Corporation, *A Primer on Solid Waste, Mobil World*, May 1972, from the ExxonMobil Historical Collection, available at the Briscoe Center for American History, University of Texas at Austin, as quoted in Sierra Club, Surfrider Foundation, Heal the Bay, and Baykeeper v. ExxonMobil Corporation, Complaint For: Nuisance; and Violation of California Unfair Competition Law, Superior Court of the State of California, County of San Francisco, September 23, 2024, https://climatecasechart.com/wp-content/uploads/case-documents/2024/20240923_docket-CGC24618321_complaint.pdf.

91 "Recycling is not . . . a government entity": People of the State of California v. ExxonMobil.

91 "someday this may be" . . . "outweigh the environmental gain": Noel Malone, manager

NOTES

of Plastics Solid Waste Management, Eastman Chemical Company, and Rolf Buhl, European Vinyls Corp., as quoted in Allen et al., *The Fraud of Plastic Recycling*.

92 "If you are a regular" ... "and it cost less": *"Plastics Recycling: Problems and Possibilities," Hearing Before the Subcommittee on Environment and Employment of the Committee on Small Business,* House of Representatives, One Hundred Second Congress, Second Session, Washington, DC, February 25, 1991, https://books.google.co.uk/books?id=xeXErBBNSV4C&printsec=frontcover&source=gbs_ge_summary_r&cad=0#v=onepage&q&f=false, 8, 120–21, 125–29.

92 "Virgin supplies" ... "kick the shit out of": Bailey Condrey, in Allen et al., *The Fraud of Plastic Recycling*, 18.

92 "They don't want to see it succeed": Tom Rattray, as quoted in Elisabeth M. Kirschner, "Recycling's Rough Adolescence," *Chemical and Engineering News* 20, November 4, 1996, https://p2infohouse.org/ref/30/29449.pdf, cited in Allen et al., *The Fraud of Plastic Recycling*, 9.

93 "the plastics industry was made" ... "recycle or be banned": William Carroll of Occidental Chemical, in *"Plastics Recycling: Problems and Possibilities,"* 121.

93 "The image of plastics" ... "beginning to take notice": Laura Sullivan et al., "Plastic Wars," *Frontline*, PBS, 2020, https://www.pbs.org/wgbh/frontline/documentary/plastic-wars/transcript/; and Sullivan et al., "Plastic Wars," at 21:15, https://www.pbs.org/wgbh/frontline/documentary/plastic-wars/.

93 "if the public thinks ... about the environment": Laura Sullivan, "How Big Oil Misled the Public into Believing Plastic Would Be Recycled," *Morning Edition*, NPR, September 11, 2020, https://www.npr.org/2020/09/11/897692090/how-big-oil-misled-the-public-into-believing-plastic-would-be-recycled.

93 "As recycling programs ... our products to grow": *Mobil Chemical Today*, undated, approximately 1993, as quoted in Sierra Club, Surfrider Foundation, Heal the Bay, and Baykeeper v. ExxonMobil.

93 was a "guilt eraser" ... "feel better about it": Freinkel, *Plastic*, 162.

93 "No doubt about it" ... their recycling programs: Wayne Pearson, executive director of the Plastics Recycling Foundation, as quoted in Myra Klockenbrink, "Plastics Industry, Under Pressure, Begins to Invest in Recycling," *New York Times*, August 30, 1988, https://www.nytimes.com/1988/08/30/science/plastics-industry-under-pressure-begins-to-invest-in-recycling.html.

94 "divert public attention away" ... era of corporate greenwashing: Rogers, *Gone Tomorrow*, 158.

96 called a "strike force": Roger Bernstein, as quoted in Freinkel, *Plastic*, 162.

96 headed to Minnesota ... would never be enforced: Sullivan et al., "Plastic Wars"; Curtis Gilbert, "Didn't Minneapolis Ban Foam Packaging Decades Ago?," *MPR News*, April 16, 2014, https://www.mprnews.org/story/2014/04/16/didnt-minneapolis-ban-foam-packaging-decades-ago; and William E. Schmidt, "Local Laws Take Aim at Indestructible Trash," *New York Times*, April 23, 1989, https://www.nytimes.com/1989/04/23/weekinreview/the-nation-local-laws-take-aim-at-indestructible-trash.html.

96 "We need to get out at the grass roots level & do guerilla warfare like our adversaries": Bailey Condrey, "1/12/94 Meeting with APME," as quoted in Allen et al., *The Fraud of Plastic Recycling*.

NOTES

96 Companies' "attitude was . . . 'pay for it'": Ronald Liesemer, in Sullivan et al., "Plastic Wars."

97 "performative investments" . . . trumpeted the work in ads: Allen et al., *The Fraud of Plastic Recycling*.

97 within five years . . . "buy recycled products": People of the State of California v. ExxonMobil.

97 Some thought 10 percent . . . "general public": Keith Atkins, former business director of solid waste management, Union Carbide Corp., as quoted in Steve Toloken, "Lofty Recycling Goals Fall by the Wayside," *Plastics News*, March 8, 1999, https://www.plasticsnews.com/article/19990308/NEWS/303089935/lofty-recycling-goals-fall-by-the-wayside.

98 "We are committed . . . the results": Irwin Levowitz, vice president, Exxon Chemical, in January 1994 meeting with APC staff, as quoted in Allen et al., *The Fraud of Plastic Recycling*, 21.

98 "provide visibility" for recycling: Krause, "An Overview Presentation."

98 bought a million-dollar: "Looking Back at Key Industry Events of '89 . . . ," *Plastics News*, March 8, 2004, https://www.plasticsnews.com/article/20040308/NEWS/303089965/looking-back-at-key-industry-events-of-89.

98 "The Urgent Need" . . . "a last resort": "The Urgent Need to Recycle," special advertising section sponsored by the Council for Solid Waste Solutions, *Time*, July 17, 1989, https://archive.org/details/time-1989-05-22/Time%201989-07-17/page/n15/mode/2up, 17–19.

99 "even fewer plastic grocery sacks" . . . "VCR tape cassettes": Mobil, "Recycling Momentum Grows," *Los Angeles Times*, 1990; Mobil, "Plastics and Recycling: Debunking a Myth," *New York Times*, 1989; NAPCOR (National Association for Plastic Container Recovery), "The Lessons of Chicken Little: A Story for Our Time," *State Legislatures*, October 1994; and American Plastics Council, "Take Another Look at Plastic," *Better Homes & Gardens*, March 1993, all as quoted in Allen et al., *The Fraud of Plastic Recycling*, 14, 44–46.

99 $18 million campaign: John Holusha, "Who Foots the Bill for Recycling?," *New York Times*, April 25, 1993, https://www.nytimes.com/1993/04/25/business/who-foots-the-bill-for-recycling.html.

99 paid $110,000 . . . might not be available everywhere: "Plastic Industry Settles Recycling Suit," United Press International, December 20, 1995, https://www.upi.com/Archives/1995/12/20/Plastic-industry-settles-recycling-suit/3518819435600/.

99 Plastic Bag Association's . . . "It is propaganda": Allen et al., *The Fraud of Plastic Recycling*, 15, 57–59.

100 "elevated to the status" . . . "clean conscience": Susan Strasser, *Waste and Want: A Social History of Trash* (Henry Holt, 1999), 285.

100 "The magnetic, gravitational power" . . . "anything else": David Allaway, senior policy analyst at the Oregon Department of Environmental Quality, as quoted in Yoder, "How the Recycling Symbol Lost Its Meaning."

100 half of states . . . skyrocketing: Elmore, *Citizen Coke*, 258.

100 "this thing just becomes everywhere": Finis Dunaway, Trent University history professor, as quoted in Yoder, "How the Recycling Symbol Lost Its Meaning."

100 polling found . . . a few years earlier: Allen et al., *The Fraud of Plastic Recycling*.

NOTES

101 The public was handing ... "decades to construct": Elmore, *Citizen Coke*, 257–60.

101 "The environmental pressure" ... "singled out": Patrick Duke, vice president of polymer service at DeWitt & Co., as quoted in Kirschner, "Recycling's Rough Adolescence"; and Bernstein, as quoted in Kirschner, "Recycling's Rough Adolescence."

101 "anti-packaging forces ... in retreat": Analyst from an Ohio-based packaging industry research firm, as quoted in Allen et al., *The Fraud of Plastic Recycling*, 20.

101 The American Plastics Council backed ... "has progressed beyond": Council spokeswoman Susan Moore and president Red Cavaney, as quoted in Roger King and Tom Ford, "APC Retreats from Goal to Recycle 25%," *Plastics News*, March 25, 1996, https://www.plasticsnews.com/article/19960325/NEWS/303259995/apc-retreats-from-goal-to-recycle-25.

102 "HIGHLY SENSITIVE POLITICALLY": Condrey notes, as quoted in Allen et al., *The Fraud of Plastic Recycling*, 14.

102 $1,500 a ton ... "reasonable financial return": Frank Aronhalt, DuPont director of environmental affairs, as quoted in Holusha, "Who Foots the Bill for Recycling?"

102 facility in the Bay Area survived barely a year: Dianne Dumanoski, "Key Events of 1996," *Plastics News*, April 23, 2004, https://www.plasticsnews.com/article/20040423/NEWS/304239998/key-events-of-1996.

102 an Oregon recycling company ... from the Plastics Council: Garten Services in Salem, Oregon, as quoted in Sullivan et al., "Plastic Wars."

102 One investigation found ... within a few years: Sullivan, "How Big Oil Misled the Public."

102 announced but never built: Kirschner, "Recycling's Rough Adolescence."

102 "recycling appears to be jinxed": Dumanoski, "Key Events of 1996."

102 "Nobody that is producing ... to replace it": Sullivan, "How Big Oil Misled the Public."

102 A Milwaukee plastic recycler ... "goals met": Marty Forman, of Poly-Anna Plastics Products, as quoted in "Dull Recycling Debate Hides Crucial Issues," *Plastics News*, February 15, 2002, https://www.plasticsnews.com/article/20020215/OPINION01/302159997/dull-recycling-debate-hides-crucial-issues; and Toloken, "Lofty Recycling Goals."

103 "of limited practicality": "Solid Waste Fact Sheet," Vinyl Institute, July 18, 1986, in Chowkwanyun, Markowitz, and Rosner, "Toxic Docs," https://cdn.toxicdocs.org/6w/6wr0N7GOdVw85VaozkQqZp3M9/6wr0N7GOdVw85VaozkQqZp3M9.pdf.

103 Its presence would clearly ... programs as a whole: Allen et al., *The Fraud of Plastic Recycling*.

103 "as an alternative to more stringent legislation": Council on Plastics and Packaging in the Environment's newsletter, as quoted in Rogers, *Gone Tomorrow*, 173.

103 bans on disposable items ... printed on it: People of the State of California v. Exxon-Mobil.

104 San Diego recycling center operator ... sell more plastic: Sullivan et al., "Plastic Wars"; and Sullivan, "How Big Oil Misled the Public."

105 China's plastic scrap imports ... by 2015: Amy L. Brooks, Shunli Wang, and Jenna R. Jambeck, "The Chinese Import Ban and Its Impact on Global Plastic Waste Trade," *Science Advances* 4, no. 6 (June 2018): figure 3, https://doi.org/10.1126/sciadv.aat0131.

NOTES

105 **1,500 shipping containers:** Will Flower, "What Operation Green Fence Has Meant for Recycling," *Waste 360*, February 11, 2016, https://www.waste360.com/waste-management-business/what-operation-green-fence-has-meant-for-recycling.

105 **Europe was exporting half . . . to China:** European Commission, *A European Strategy for Plastics in a Circular Economy*, 16, https://www.europarc.org/wp-content/uploads/2018/01/Eu-plastics-strategy-brochure.pdf.

105 **In Wen'an . . . lung ailments and other illnesses:** Franklin-Wallis, *Wasteland*, 76–79.

106 **"Those folks were getting so screwed financially":** California state Senator Ben Allen, interview with author, July 9, 2024.

107 **9 percent of the world's plastic waste has ever been recycled:** Roland Geyer, Jenna R. Jambeck, and Kara Lavender Law, "Production, Use, and Fate of All Plastics Ever Made," *Science Advances* 3, no. 7 (July 2017), https://doi.org/10.1126/sciadv.1700782.

107 **below 6 percent:** *The Real Truth About the U.S. Plastics Recycling Rate* (Beyond Plastics and the Last Beach Cleanup, May 4, 2022), https://static1.squarespace.com/static/5eda91260bbb7e7a4bf528d8/t/62b2238152acae761414d698/1655841666913/The-Real-Truth-about-the-US-Plastic-Recycling-Rate-2021-Facts-and-Figures-_5-4-22.pdf.

107 **Australia, for example, reports recycling about 14 percent:** Department of Climate Change, Energy, the Environment and Water, "Australian Plastic Flows and Fates Study 2021–22 National Report," Australian Government, 2024, https://www.dcceew.gov.au/environment/protection/waste/publications/australian-plastic-flows-and-fates-national-report-2021-22.

107 **an independent analysis . . . by a third:** "UK Overestimates Plastic Recycling by a Third," Eunomia, March 6, 2018, https://eunomia.eco/uk-overestimates-plastic-recycling-by-a-third/.

108 **"It used to be" . . . "happening now":** ExxonMobil Baytown Area (@ExxonMobilBTA), "It used to be that only SOME plastics could be recycled. Now, thanks to our revolutionary advanced recycling technology, we can give any plastic a new life. A chip bag into a Ferrari? A water bottle into a computer? Is it Science Fiction? It's happening now at ExxonMobil Baytown," Twitter (now X), August 28, 2023, https://x.com/exxonmobilbta/status/1696261899652931725?s=46&t=OBruA2TmyQn2AZvSTY+mlOQ.

108 **"there are no evident . . . advanced recycling processes":** Karen McKee, "ExxonMobil Steps Up Advanced Recycling to Help Address Plastic Waste," ExxonMobil, March 30, 2021, https://corporate.exxonmobil.com/news/viewpoints/steps-up-advanced-recycling-plastic-waste.

108 **Very little of what . . . new plastic; naphtha from waste . . . less than 5 percent:** Lisa Song, "Selling a Mirage," *ProPublica*, June 20, 2024, https://www.propublica.org/article/delusion-advanced-chemical-plastic-recycling-pyrolysis.

108 **At the plant ExxonMobil's gushy tweet referenced, it's just 8 percent:** People of the State of California v. ExxonMobil.

109 **One government study . . . 1 percent of the waste plastic put in:** Taylor Uekert et al., "Technical, Economic, and Environmental Comparison of Closed-Loop Recycling Technologies for Common Plastics," *ACS Sustainable Chemistry and Engineering* 11 (January 2023): 965–78, 969, https://doi.org/10.1021/acssuschemeng.2c05497.

109 **at most .09 percent plastic waste . . . attorney general:** People of the State of California v. ExxonMobil.

NOTES

109 "mathematical acrobatics": Song, "Selling a Mirage."

109 "the imaginary realm": People of the State of California v. ExxonMobil.

109 "blatant misstatements" . . . out of landfills: ExxonMobil Corporation v. Robert Andres Bonta a.k.a. Rob Bonta, in His Individual Capacity; Sierra Club, Inc.; Surfrider Foundation, Inc.; Heal the Bay, Inc.; Baykeeper, Inc. and Intergenerational Environment Justice Fund Ltd., United States District Court for the Eastern District of Texas, Beaumont Division, January 6, 2025, https://www.documentcloud.org/documents/25483310-010624-ed-tex-exxon-mobil-corp-v-bonta-et-al-complaint/?ref=floodlightnews.org#document/p1.

109 "putting a nail in the coffin": Tom Crotty, interview with author, September 19, 2024.

110 "a whole war" happening "behind closed doors": Rose Ní Chléirigh, assistant to European Parliament member Grace O'Sullivan, interview with author, March 20, 2024.

110 unregulated, voluntary system authorizing "certified circular polymers": International Sustainability & Carbon Certification, ISCC PLUS.

110 more carbon emissions . . . ends up incinerated: People of the State of California v. ExxonMobil.

110 "a narrative solution": Joan Marc Simon, Zero Waste Europe, interview with author March 21, 2024.

110 ExxonMobil internal communications . . . "will be invaluable": People of the State of California v. ExxonMobil.

111 industry is paying TikTok and Instagram influencers . . . "eliminate plastic waste": Hiroko Tabuchi, "Inside the Plastic Industry's Battle to Win Over Hearts and Minds," *New York Times*, November 27, 2024, https://www.nytimes.com/2024/11/27/climate/plastic-industry-internal-documents.html.

112 One study found people . . . "encourages disposability": Franklin-Wallis, *Wasteland*, 325.

112 plans to package a quarter of drinks in reusable containers: Coca-Cola Company, *2021 World Without Waste Report*, 2, https://www.coca-colacompany.com/content/dam/company/us/en/reports/pdf/coca-cola-world-without-waste-report-2021.pdf.

112 yet more recycling promises: "The Coca-Cola Company Evolves Voluntary Environmental Goals," Coca-Cola Company, December 2, 2024, https://www.coca-colacompany.com/media-center/the-coca-cola-company-evolves-voluntary-environmental-goals.

Chapter 5: "A Fantastic Window of Opportunity"

113 most commonly used material in the world: "A History of Plastics," British Plastics Federation, accessed February 16, 2025, https://www.bpf.co.uk/plastipedia/plastics_history/Default.aspx.

113 "A Fantastic Window of Opportunity": Alexander H. Tullo, "Why the Future of Oil Is in Chemicals, Not Fuels," *Chemical and Engineering News*, February 20, 2019, https://cen.acs.org/business/petrochemicals/future-oil-chemicals-fuels/97/i8.

113 In 1930, the Standard Oil Company of California . . . into their venture: Ellen R. Wald, *Saudi, Inc.: The Arabian Kingdom's Pursuit of Profit and Power* (Pegasus Books, 2018), 7–20.

114 For Aramco . . . around $2 billion: Wald, *Saudi, Inc.*, 54–75, 183, 207.

115 pearl diver's son: Stanley Reed, "An Oil Giant Is Taking Big Steps. Saudi Arabia Can't Afford

NOTES

for It to Slip," *New York Times,* June 16, 2018, https://www.nytimes.com/2018/06/16/business/energy-environment/saudi-arabia-aramco.html.

115 **born three years before . . . "guarantee a market":** Wald, *Saudi, Inc.*, 111, 221, 236–37.

116 **surpassed Apple to become the world's most valuable company:** Graeme Wearden, "Saudi Aramco Overtakes Apple as World's Most Valuable Company," *The Guardian*, May 12, 2022, https://www.theguardian.com/business/2022/may/12/saudi-aramco-overtakes-apple-worlds-most-valuable-company.

116 **third-biggest by revenue, raking in nearly $500 billion a year:** "Companies Ranked by Revenue," CompaniesMarketCap, accessed April 11, 2025, https://companiesmarketcap.com/gbp/largest-companies-by-revenue.

116 **nearly $1.7 trillion—almost four times as much as ExxonMobil:** "Largest Companies by Market Cap," CompaniesMarketCap, accessed April 11, 2025, https://companiesmarketcap.com/.

116 **two-thirds of government revenue:** Jack Dutton, "Saudi Arabia 'Conservative' on Oil Revenue as It Pares Back Vision 2030 Plans," *Al-Monitor*, May 14, 2024, https://www.al-monitor.com/originals/2024/05/saudi-arabia-conservative-oil-revenue-it-pares-back-vision-2030-plans#:~:text=Saudi%20Arabia%20is%20currently%20the,kingdom's%20revenues%20come%20from%20oil.

116 **"Protecting the business" . . . "ruling family":** Jim Krane, "The Bottom of the Barrel: Saudi Aramco and Global Climate Action," working paper, 2021, Rice University's Baker Institute for Public Policy, https://www.bakerinstitute.org/sites/default/files/2021-01/import/ces-wp-saudiaramco-010821.pdf.

116 **"more of a threat than climate change itself":** Joanna Depledge, Kari De Pryck, and J. Timmons Roberts, "Decades of Systematic Obstructionism: Saudi Arabia's Role in Slowing Progress in UN Climate Negotiations," Climate Social Science Network, November 2023, https://cssn.org/wp-content/uploads/2023/11/Decades-of-Systematic-Obstructionism-CSSN-Issue-Paper3.pdf.

116 **An analysis of history's . . . former Soviet Union:** "Carbon Majors Entities," InfluenceMap and Carbon Majors, April 2024, https://carbonmajors.org/Entities.

117 **"We only produce oil because people want to buy it":** Olivier Thorel, Aramco's vice president for chemicals and hydrogen, as quoted in Tom Wilson, "Saudi Aramco Bets on Being the Last Oil Major Standing," *Financial Times*, January 12, 2023, https://www.ft.com/content/513b770b-836b-472b-a058-3e4a95437c69.

117 **"the most profitable and the most polluting company of all time":** "Market Debut for World's Biggest Polluter Must Be a Rallying Cry for Climate Action," *The Guardian*, December 8, 2019, https://www.theguardian.com/business/2019/dec/08/saudi-aramco-flotation-biggest-polluter-climate-action.

117 **"Environmental stewardship" . . . coastal mangroves:** Ashraf Al Ghazzawi, Aramco's executive vice president for strategy, as quoted in Stanley Reed, "Saudi Arabia Eyes a Future Beyond Oil," *New York Times*, May 29, 2024, https://www.nytimes.com/2024/05/29/business/saudi-arabia-renewable-energy-solar-wind.html.

117 **"for generations to come":** Saudi Aramco 2019 bond prospectus, as quoted in Krane, "Bottom of the Barrel."

117 **"thirty-year record" . . . "paralyzing discussions":** Depledge, De Pryck, and Roberts, "Systematic Obstructionism."

NOTES

118 **Other nations at . . . "'La La Land'":** Linda Kalcher, a former climate adviser to the United Nations, and Prince Abdulaziz bin Salman, as quoted in Lisa Friedman, Brad Plumer, and Vivian Nereim, "Saudi Arabia Is Trying to Block a Global Deal to End Fossil Fuels, Negotiators Say," *New York Times*, December 10, 2023, https://www.nytimes.com/2023/12/10/climate/saudi-arabia-cop28-fossil-fuels.html.

118 **"The more they postpone, the more they earn":** Morten Kaldhussæter Flisnes, "Where You Stand Depends on What You Sell: Saudi Arabia's Obstructionism in the UNFCCC 2012–2018," Center for International Climate Research (CICERO), August 29, 2019, https://pub.cicero.oslo.no/cicero-xmlui/bitstream/handle/11250/2611866/Rapport%202019%2011%20-%20web.pdf?sequence=1.

118 **as recently as 2022 . . . UN document:** Friedman, Plumer, and Nereim, "Saudi Arabia Is Trying to Block a Global Deal."

118 **"keep the world hooked" . . . boosting gasoline-powered ones:** Hiroko Tabuchi, "Inside the Saudi Strategy to Keep the World Hooked on Oil," *New York Times*, November 21, 2022, https://www.nytimes.com/2022/11/21/climate/saudi-arabia-aramco-oil-solar-climate.html.

118 **Across Africa . . . "for the kingdom":** Lawrence Carter and Tom Costello, "Cheap Cars, Supersonic Jets and Floating Power Plants: Undercover in Saudi Arabia's Secretive Program to Keep the World Burning Oil," Centre for Climate Reporting with Channel 4 News, November 27, 2023, https://climate-reporting.org/undercover-saudi-arabia-keep-burning-oil/.

119 **In 2017, Saudi Aramco . . . "every hydrocarbon molecule":** "Sadara," Saudi Arabian Oil Co., accessed September 6, 2024, https://www.aramco.com/en/what-we-do/mega-projects/sadara-petrochemicals-facility.

119 **tires and auto interiors to diapers, flooring, packaging, and appliances:** "PlasChem Park," Sadara, accessed September 6, 2024, https://www.sadara.com/en/PlasChem_Park.

119 **"a powerful statement" . . . "attached by an umbilical cord":** Anjli Raval and Andrew Ward, "Saudi Aramco Plans for a Life After Oil," *Financial Times*, December 10, 2017, https://www.ft.com/content/e46162ca-d9a6-11e7-a039-c64b1c09b482.

120 **"We see the world" . . . "keep our market share":** Abdulaziz al-Judaimi, Aramco's senior vice president for chemicals and refining, as quoted in Reed, "An Oil Giant Is Taking Big Steps."

120 **"big picture imperative" . . . "to leave reserves stranded":** John Richardson, "Details of How Saudi Aramco COTC and Other Advantaged Feedstock Projects Could Redraw the Petrochemicals Map," ICIS: Independent Commodity Intelligence Services, November 3, 2023, https://www.icis.com/asian-chemical-connections/2023/11/details-of-how-saudi-aramco-cotc-and-other-advantaged-feedstock-projects-could-redraw-the-petrochemicals-map/.

120 **fourth-biggest . . . close to $35 billion:** Alexander H. Tullo, "C&EN's Global Top 50 for 2020," *Chemical and Engineering News*, July 27, 2020, https://cen.acs.org/business/finance/CENs-Global-Top-50-2020/98/i29.

120 **"one of Saudi Arabia's crown jewels":** Aaron Schaffer, "Edelman Strikes $6 Million Deal with Saudis to Promote G-20 Business Summit," *Al-Monitor*, February 19, 2020, https://www.al-monitor.com/originals/2020/02/sabic-g20-b20-saudi-arabia-edelman-public-relations-hire.html.

NOTES

120 **GE Plastics . . . SABIC paid $11.6 billion in 2007**: "GE Announces Sale of Plastics Business to SABIC for $11.6 Billion," GE, May 25, 2007, https://www.ge.com/news/press-releases/ge-announces-sale-plastics-business-sabic-116-billion-industrial-portfolio.

121 **more than three times Saudi Aramco's**: "Saudi Aramco Signs Share Purchase Agreement to Acquire 70% Majority Stake in SABIC from the Public Investment Fund of Saudi Arabia," Saudi Aramco, March 27, 2019, https://www.aramco.com/en/news-media/news/2019/aramco-sabic.

121 **"great potential" . . . construction**: "Aramco Picks Oil-to-Chemicals as Growth Engine," Gulf Petrochemicals and Chemicals Association Forum, December 5, 2018, https://www.gpcaforum.com/2018/12/05/aramco-picks-oil-to-chemicals-as-growth-engine/.

121 **nearly half the growth . . . "future oil production"**: "Aramco to Invest More Than $100bn over the Next 10 Years in Chemicals," *Saudi Gazette*, November 27, 2018, https://www.saudigazette.com.sa/article/549019/BUSINESS/Aramco-to-invest-more-than-$100bn-over-the-next-10-years-in-chemicals.

121 **"As populations grow they need more plastic"**: Abdulaziz al-Judaimi, as quoted in Rania El Gamal, "Exclusive: Saudi Aramco Eyes Partnerships as It Expands Refining, Petrochems," Reuters, June 12, 2018, https://www.reuters.com/article/us-saudi-aramco-downstream-exclusive-idUSKBN1J81M9/.

121 **provided "a fantastic window of opportunity" . . . "those who act quickly"**: Tullo, "Future of Oil."

121 **By the early 2030s . . . near-tripling in fifteen years**: Wilson, "Last Oil Major Standing."

121 **"A game changer" . . . "ever to impact"**: Mark Thomas, "Crude Oil–to–Chemicals: A Game Changer for the Chemical Industry," *Chemical Week*, November 29, 2019, https://chemweek.com/CW/Document/107708/Crude-oiltochemicals-A-game-changer-for-the-chemical-industry.

122 **heat-based technology . . . 25 percent of the oil it takes in into chemicals**: "Aramco Affiliate S-OIL to Build One of the World's Largest Petrochemical Crackers in South Korea," Saudi Arabian Oil Co., November 17, 2022, https://www.aramco.com/en/news-media/news/2022/aramco-affiliate-s-oil-to-build-one-of-the-worlds-largest-petrochemical-crackers-in-south-korea.

122 **"Definitely a breakthrough"**: Richardson, "Details."

122 **"What if the yield" . . . "hitting the jackpot"**: "Crude Oil to Chemicals," Saudi Arabian Oil Co., accessed September 6, 2024, https://www.aramco.com/en/what-we-do/energy-innovation/advancing-energy-solutions/crude-oil-to-chemicals.

123 **"easily" push . . . "global supply and demand"**: R. J. Chang, vice president of Process Economics Program at IHS Markit, as quoted in Tullo, "Future of Oil."

123 **If Saudi Aramco reaches . . . "reshape the global petrochemical industry"**: R. J. Chang and Dewey Johnson, "Crude Oil-to-Chemicals Projects Presage a New Era in Global Petrochemical Industry," S&P Global, August 6, 2018, https://web.archive.org/web/20241008145740/https://www.spglobal.com/commodityinsights/en/ci/research-analysis/crudeoil-chemicals-projects.html.

123 **a big crude-to-chemicals facility beside the Persian Gulf and possibly another on the Red Sea**: Located, and potentially located, in Jubail and Yanbu.

123 **could double . . . per barrel**: Thomas, "Game Changer."

NOTES

123 **more than four million tons of ethylene:** Richardson, "Details."

124 **"must fundamentally rethink":** Tim Fitzgibbon et al., "From Crude Oil to Chemicals: How Refineries Can Adapt to Shifting Demand," McKinsey & Company, June 30, 2022, https://www.mckinsey.com/industries/chemicals/our-insights/from-crude-oil-to-chemicals-how-refineries-can-adapt-to-shifting-demand.

124 **"Any time we look"... "haven't seen before":** Glover, as quoted in Tullo, "Future of Oil."

126 **seventh-biggest corporate plastic polluter:** "The Brand Audit 2023 Report," Break Free from Plastic, accessed September 21, 2024, https://brandaudit.breakfreefromplastic.org/brand-audit-2023/#:~:text=The%20Brand%20Audit%20Report%202023,burning%20plastic%20and%20chemical%20recycling.

126 **"creating products"... "to achieve this":** Turki Binmoammar, managing director, Saudi Arabia government relations, and public policy senior director, Procter & Gamble Middle East, speaking at the Gulf Petrochemicals & Chemicals Association's 13th GPCA Plastics Conference, Dubai, United Arab Emirates, May 13–14, 2024.

126 **"to protect the climate... close the loop":** Rania Zeidan, senior account manager, packaging and specialty plastics, Dow Chemical Company, speaking at the 13th GPCA Plastics Conference.

126 **"very, very important"... "a problem":** Priya Sarma, head of sustainability and corporate affairs, Unilever Middle East, Turkey, speaking at the 13th GPCA Plastics Conference.

131 **Over five years, it has... Europe, Japan, and Korea combined:** Cierán Healy, "China's Petrochemical Surge Is Driving Global Oil Demand Growth," International Energy Agency, December 19, 2023, https://www.iea.org/commentaries/china-s-petrochemical-surge-is-driving-global-oil-demand-growth.

131 **China nearly doubled... polyester and PET:** Reuters Graphics, "China's Petrochemical Demand Is Rising Fast," Reuters, accessed October 21, 2024, https://fingfx.thomsonreuters.com/gfx/editorcharts/CHINA-PETROCHEMICALS/0H0014BT121Y/index.html.

131 **"dwarfs any historical precedent," roughly doubling:** Healy, "China's Petrochemical Surge."

131 **On Dayushan Island... sixteen million tons:** Alexander H. Tullo, "China's Aromatics Building Boom Rattles the Petrochemical Industry," *Chemical and Engineering News*, March 12, 2024, https://cen.acs.org/business/petrochemicals/China-aromatics-building-boom-rattles-the-petrochemical-industry/102/i8.

131 **"need to rethink their business model":** Khalid Al Dawood, managing director at National Petrochemical Industrial Company, speaking at the 13th GPCA Plastics Conference.

132 **In late 2023... "have been possible":** Healy, "China's Petrochemical Surge."

Chapter 6: Invisible Poisons

133 **Theo Colborn was always... mothers took decades earlier:** Theo Colborn, John Peterson Myers, and Dianne Dumanoski, *Our Stolen Future: Are We Threatening Our Fertility, Intelligence and Survival? A Scientific Detective Story* (Little, Brown, 1996), 11–28, 46–48, 51–67.

134 **one drop in twenty Olympic swimming pools:** Marsha Richmond, "The Endocrine

NOTES

Disruptor Exchange: Exploring Environmental Chemicals," Theo Colborn and Endocrine Disruption, accessed February 6, 2025, https://theocolborn.com/tedx-the-endocrine-disruptor-exchange-2000-2020/.

136 **Those exposed in the womb... span three generations:** Carrie Arnold, "Consequences of DDT Exposure Could Last Generations," *Scientific American*, July 1, 2021, https://www.scientificamerican.com/article/consequences-of-ddt-exposure-could-last-generations/.

137 **"is a lock"... 6 percent of the suspects:** University of Missouri biologist Frederick vom Saal and Bengt-Erik Bengtsson, of the Swedish Environmental Protection board, as quoted in Colborn, Myers, and Dumanoski, *Our Stolen Future*, 40, 75, 17–18.

138 **"like playing a game with molecules":** "Plastics Additives," British Plastics Federation, accessed February 6, 2025, https://www.bpf.co.uk/plastipedia/additives/Default.aspx.

138 **nearly half of a finished plastic's weight:** Ksenia J. Groh et al., "Overview of Known Plastic Packaging-Associated Chemicals and Their Hazards," *Science of the Total Environment* 651 (February 2019): 3253–68, https://doi.org/10.1016/j.scitotenv.2018.10.015.

138 **more than sixty individual chemicals:** Martin Wagner et al., "State of the Science on Plastic Chemicals—Identifying and Addressing Chemicals and Polymers of Concern," Plast-Chem Project, 2024, 30, https://zenodo.org/records/15397723.

139 **among the world's most widely used synthetic chemicals:** "List of Endocrine Disruptors: The Not-So-Happy Families of Toxic Chemicals," CHEM Trust, accessed February 6, 2025, https://chemtrust.org/edcs-list/.

139 **seven billion kilograms... annually:** Meg Wilcox, "How BPA and Its Evil Cousins Dodge Meaningful Regulation," *Environmental Health News*, January 8, 2024, https://www.ehn.org/bpa-regulations-2666462524.html.

139 **"large scale dysfunction at the population level is possible":** Colborn, Myers, and Dumanoski, *Our Stolen Future*, 252–54.

140 **now well documented:** Hagai Levine et al., "Temporal Trends in Sperm Count: A Systematic Review and Meta-Regression Analysis of Samples Collected Globally in the 20th and 21st Centuries," *Human Reproduction Update* 29, no. 2 (March 2023): 157–76, https://doi.org/10.1093/humupd/dmac035.

140 **"the danger of widespread"... "a sequel to *Silent Spring*":** Colborn, Myers, and Dumanoski, *Our Stolen Future*, v–vii, 81, 140, 165, 170–71, 202–7.

141 **"one part pseudo-science... three parts hysteria":** Ronald Bailey, "Hormones and Humbug," *Washington Post*, March 31, 1996, https://www.washingtonpost.com/archive/opinions/1996/03/31/hormones-and-humbug/142b18c2-b362-4ca7-88d9-71156f67be26/.

141 **sent reporters... such a response:** Rick Weiss and Gary Lee, "Pollution's Effect on Human Hormones: When Fear Exceeds Evidence," *Washington Post*, March 31, 1996, https://www.washingtonpost.com/archive/politics/1996/03/31/pollutions-effect-on-human-hormones-when-fear-exceeds-evidence/05eb0577-cebc-4477-8754-f6efb13a50e1/.

142 **she had seen in her days at *The Boston Globe*... "wonderful alternative term":** "Endocrine Disruption and Environmental Health: Ten Years After *Our Stolen Future*," transcript of a call held by the Collaborative for Health and Environment, March 22, 2006, https://www.healthandenvironment.org/partnership_calls/346.

142 **sign her emails "Onward!":** Elizabeth Grossman et al., "Theodora (Theo) Colborn: 1927–2014," *Environmental Health Perspectives* 123, no. 3 (March 2015), https://doi.org/10.1289/ehp.1509743.

NOTES

142 "we are in far deeper" . . . "hundreds, now": "Endocrine Disruption and Environmental Health," call transcript.

143 the vast majority of people tested: Philip J. Landrigan et al., "The Minderoo-Monaco Commission on Plastics and Human Health," *Annals of Global Health* 89, no. 1 (2023), https://doi.org/10.5334/aogh.4056; and "Human Exposure to Bisphenol A in Europe," European Environment Agency, September 14, 2024, https://www.eea.europa.eu/publications/peoples-exposure-to-bisphenol-a.

143 premature ovarian failure . . . early menopause: A. C. Gore et al., "EDC-2: The Endocrine Society's Second Scientific Statement on Endocrine-Disrupting Chemicals," *Endocrine Reviews* 36, no. 6 (December 2015): E1–150, https://doi.org/10.1210/er.2015-1010.

143 phthalate exposure is correlated with endometriosis: Huan Yi et al., "Phthalate Exposure and Risk of Ovarian Dysfunction in Endometriosis: Human and Animal Data," *Frontiers in Cell and Developmental Biology* 11 (July 2023), https://doi.org/10.3389/fcell.2023.1154923.

143 were 60 percent more likely to miscarry than those with the lowest: "Exposure to Phthalates May Raise Risk of Pregnancy Loss, Gestational Diabetes," Harvard T. H. Chan School of Public Health, November 8, 2016, https://www.hsph.harvard.edu/news/features/phthalates-exposure-pregnancy-loss-gestational-diabetes/.

143 near-tripling of the likelihood . . . early deliveries a year: Leonardo Trasande et al., "Prenatal Phthalate Exposure and Adverse Birth Outcomes in the USA: A Prospective Analysis of Births and Estimates of Attributable Burden and Costs," *The Lancet Planetary Health* 8, no. 2 (February 2024): e74–85, https://doi.org/10.1016/s2542-5196(23)00270-x.

143 may even change its shape and size: Yuan-duo Zhu et al., "Prenatal Phthalate Exposure and Placental Size and Shape at Birth: A Birth Cohort Study," *Environmental Research* 160 (January 2018): 239–46, https://doi.org/10.1016/j.envres.2017.09.012.

143 undescended testicles . . . lower fertility: Landrigan et al., "The Minderoo-Monaco Commission."

143 decreased sperm counts and lower testosterone levels: Elizabeth G. Radke et al., "Phthalate Exposure and Male Reproductive Outcomes: A Systematic Review of the Human Epidemiological Evidence," *Environment International* 121 (December 2018): 764–93, https://doi.org/10.1016/j.envint.2018.07.029.

144 highest levels of BPA . . . higher chance of a stroke: Shaofang Cai et al., "Relationship Between Urinary Bisphenol A Levels and Cardiovascular Diseases in the U.S. Adult Population, 2003–2014," *Ecotoxicology and Environmental Safety* 192 (April 2020): 110300, https://doi.org/10.1016/j.ecoenv.2020.110300.

144 may contribute to kidney disease: Rafael Moreno-Gómez-Toledano et al., "Bisphenol A Exposure and Kidney Diseases: Systematic Review, Meta-Analysis, and NHANES 03–16 Study," *Biomolecules* 11, no. 7 (July 2021): 1046, https://doi.org/10.3390/biom11071046.

144 interfere with the body's metabolism . . . insulin resistance later: Gore et al., "EDC-2."

144 breast, uterus, ovaries, and prostate: Endocrine Society, "Impact of EDCs on Hormone-Sensitive Cancer," accessed February 7, 2025, https://www.endocrine.org/topics/edc/what-edcs-are/common-edcs/cancer.

144 DEHP, a common ingredient . . . promotes the cancer's spread: Douglas Fischer, "Plastic Additive Increases Breast Cancer Relapse, Mortality: New Science," *Environmental Health News*, December 2, 2021, https://www.ehn.org/dehp-breast-cancer-2655803837.html.

NOTES

144 "This may be the tip of the iceberg": Pete Myers interview with Bobby Bascomb and Steve Curwood, in "Medical Plastic Linked to Breast Cancer Relapse," *Living on Earth*, December 10, 2021, https://www.loe.org/shows/segments.html?programID=21-P13-00050&segmentID=2.

144 ADHD and autism spectrum disorder: Stephanie M. Engel et al., "Prenatal Phthalates, Maternal Thyroid Function, and Risk of Attention-Deficit Hyperactivity Disorder in the Norwegian Mother and Child Cohort," *Environmental Health Perspectives* 126, no. 5 (May 2018), https://doi.org/10.1289/ehp2358; Anne-Louise Ponsonby et al., "Prenatal Phthalate Exposure, Oxidative Stress-Related Genetic Vulnerability and Early Life Neurodevelopment: A Birth Cohort Study," *NeuroToxicology* 80 (September 2020): 20–28, https://doi.org/10.1016/j.neuro.2020.05.006; Youssef Oulhote et al., "Gestational Exposures to Phthalates and Folic Acid, and Autistic Traits in Canadian Children," *Environmental Health Perspectives* 128, no. 2 (February 2020), https://doi.org/10.1289/ehp5621; and T. Peter Stein et al., "Bisphenol A Exposure in Children with Autism Spectrum Disorders," *Autism Research* 8, no. 3 (June 2015): 272–83, https://doi.org/10.1002/aur.1444.

144 lower IQ, behavioral problems, and reduced motor and communication skills: Maede Ejaredar et al., "Phthalate Exposure and Children's Neurodevelopment: A Systematic Review," *Environmental Research* 142 (October 2015): 51–60, https://doi.org/10.1016/j.envres.2015.06.014; Dong-Wook Lee et al., "Prenatal and Postnatal Exposure to Di-(2-Ethylhexyl) Phthalate and Neurodevelopmental Outcomes: A Systematic Review and Meta-Analysis," *Environmental Research* 167 (November 2018): 558–66, https://doi.org/10.1016/j.envres.2018.08.023.

144 "a neuropsychiatric problem" . . . Parkinson's diseases: Elizabeth Ryznar, "The Plastics Crisis: A Neuropsychiatric Problem Hidden in Plain Sight," *Psychiatric Times*, September 3, 2024, https://www.psychiatrictimes.com/view/the-plastics-crisis-a-neuropsychiatric-problem-hidden-in-plain-sight.

145 "doesn't seem to have" . . . "safe level": Vom Saal, as quoted in Wilcox, "BPA and Its Evil Cousins."

145 one analysis attributed . . . the phthalate DEHP: Sara Hyman et al., "Phthalate Exposure from Plastics and Cardiovascular Disease: Global Estimates of Attributable Mortality and Years Life Lost," *eBioMedicine*, April 28, 2025, 105730, https://doi.org/10.1016/j.ebiom.2025.105730.

145 Another study estimated . . . US alone: Landrigan et al., "The Minderoo-Monaco Commission."

146 "a bright, bold line" . . . "entire life span": Leo Trasande, New York University pediatrician, as quoted in Susanne Rust, "More Concerning Than the Nanoplastics in Water Bottles Are the Chemicals on Them," *Los Angeles Times*, January 18, 2024, htps://www.latimes.com/environment/story/2024-01-18/health-costs-of-plastics-run-250-billion-a-year.

146 more than ten thousand, and said . . . harmful to human health: Helene Wiesinger, Zhanyun Wang, and Stefanie Hellweg, "Deep Dive into Plastic Monomers, Additives, and Processing Aids," *Environmental Science and Technology* 55, no. 13 (June 2021): 9339–51, https://doi.org/10.1021/acs.est.1c00976.

146 thousands of new chemicals: Endocrine Society, "Endocrine-Disrupting Chemicals," https://www.endocrine.org/-/media/endocrine/files/advocacy/position-statement/position_statement_endocrine_disrupting_chemicals.pdf.

NOTES

146 **"It was written" ... "keep it from doing much":** "The Toxic Substances Control Act: From the Perspective of Steven D. Jellinek," Science History Institute Museum and Library, interview by Jody A. Roberts and Kavita D. Hardy at the Chemical Heritage Foundation, Philadelphia, Pennsylvania, January 29, 2010 (Philadelphia: Chemical Heritage Foundation, Oral History Transcript no. 0653), https://digital.sciencehistory.org/works/jvt64zj.

148 **"protest everything" approach:** Judge Jerry E. Smith, US Court of Appeals, Fifth Circuit, denial of rehearing, November 27, 1991, Corrosion Proof Fittings, et al. v. the Environmental Protection Agency and William K. Reilly, https://openjurist.org/947/f2d/1201/corrosion-proof-fittings-v-environmental-protection-agency.

148 **Fear of lawsuits ... "just hasn't happened":** "The Toxic Substances Control Act: From the Perspective of Steven D. Jellinek."

148 **Boxer said ... denied Boxer's claim:** "Senator Boxer's Statement for Press Conference on Udall-Vitter TSCA Bill," US Senate Committee on Environment and Public Works, March 17, 2015, https://www.epw.senate.gov/public/index.cfm/2015/3/post-9ff1060b-e511-b1c6-83a3-9c8efe467cae; and Neil Bedi, Sharon Lerner, and Kathleen McGrory, "Why the U.S. Is Losing the Fight to Ban Toxic Chemicals," *ProPublica*, December 14, 2022, https://www.propublica.org/article/toxic-chemicals-epa-regulation-failures.

149 **"emptying the sea with a teaspoon":** Olatz Fínex Marañon et al., *From Risk to Resilience: Navigating Towards a Toxic-Free Future* (European Environmental Bureau, April 2024), 5, https://eeb.org/wp-content/uploads/2024/04/From-Risk-to-Resilience-Navigating-Towards-a-Toxic-Free-Future.pdf.

149 **carry their own harms:** Michael Thoene et al., "Bisphenol S in Food Causes Hormonal and Obesogenic Effects Comparable to or Worse Than Bisphenol A: A Literature Review," *Nutrients* 12, no. 2 (February 2020): 532, https://doi.org/10.3390/nu12020532; and Anna Marqueño et al., "Toxic Effects of Bisphenol A Diglycidyl Ether and Derivatives in Human Placental Cells," *Environmental Pollution* 244 (January 2019): 513–21, https://doi.org/10.1016/j.envpol.2018.10.045.

149 **chemicals used to replace ... DEHP's punch:** Trasande et al., "Prenatal Phthalate Exposure."

149 **"a never-ending shell game":** Washington State University biologist Patricia Hunt, as quoted in Wilcox, "BPA and Its Evil Cousins."

149 **"hasn't been given" ... "opposed to regulation":** Michal Freedhoff, then head of chemical regulation at EPA; Robert Sussman, former EPA deputy administrator; Joel Tickner, University of Massachusetts, Lowell environmental health professor, as quoted in Bedi, Lerner, and McGrory, "Why the U.S. Is Losing the Fight to Ban Toxic Chemicals."

150 **budget recently hit $180 million:** Andrea Suozzo et al., "Nonprofit Explorer: American Chemistry Council," *ProPublica*, accessed February 8, 2025, https://projects.propublica.org/nonprofits/organizations/530104410.

150 **first half of the 2020s, as the Biden administration ... influence efforts:** "Top Spenders," OpenSecrets, 2020–2024 cycles, accessed February 8, 2025, https://www.opensecrets.org/federal-lobbying/top-spenders.

150 **Its political arm ... Republicans than Democrats:** "Chemical & Related Manufacturing PACS Contributions to Candidates, 2023–2024," OpenSecrets, accessed February 8, 2025,

NOTES

https://www.opensecrets.org/political-action-committees-pacs/industry-detail/N13/2024; and "PAC Profile: American Chemistry Council," OpenSecrets, accessed February 8, 2025, https://www.opensecrets.org/political-action-committees-pacs/american-chemistry-council/C00252338/summary/2024.

150 **says it prioritizes . . . "available science"**: "TSCA 1,4-Dioxane Risk Evaluation Is Not Consistent with the State of the Science or Assessments from Other Regulatory Bodies," American Chemistry Council, November 14, 2024, https://www.americanchemistry.com/chemistry-in-america/news-trends/press-release/2024/tsca-1-4-dioxane-risk-evaluation-is-not-consistent-with-the-state-of-the-science-or-assessments-from-other-regulatory-bodies.

150 **one disclosure . . . "one after another"**: Then–state representative Diana Urban, as quoted in Allegra Abramo, "In New Battleground over Toxic Reform, American Chemistry Council Targets the States," *InvestigateWest*, September 10, 2013, https://www.investigatewest.org/investigatewestreports/in-new-battleground-over-toxic-reform-american-chemistry-council-targets-the-states-17692491.

151 **fifteen of the top twenty environment jobs**: Lisa Friedman and Claire O'Neill, "Who Controls Trump's Environmental Policy?," *New York Times*, January 14, 2020, https://www.nytimes.com/interactive/2020/01/14/climate/fossil-fuel-industry-environmental-policy.html.

151 **Nancy Beck jumped . . . "invested in a chemical"**: Eric Lipton, "Why Has the E.P.A. Shifted on Toxic Chemicals? An Industry Insider Helps Call the Shots," *New York Times*, October 21, 2017, https://www.nytimes.com/2017/10/21/us/trump-epa-chemicals-regulations.html.

151 **"recognizable steps in the right direction"**: Endocrine Society, "Endocrine-Disrupting Chemicals."

152 **"At least it's doing something"**: George Monbiot, "Britain Is Becoming a Toxic Chemical Dumping Ground—Yet Another Benefit of Brexit," *The Guardian*, March 18, 2024, https://www.theguardian.com/commentisfree/2024/mar/18/britain-toxic-chemical-dump-brexit-europe.

152 **banned or restricted the use of about two thousand chemicals**: European Environmental Bureau, "The Great Detox—Largest Ever Ban of Toxic Chemicals Announced by EU," April 25, 2022, https://eeb.org/the-great-detox-largest-ever-ban-of-toxic-chemicals-announced-by-eu/.

152 **more than a dozen phthalates . . . agency estimated**: "Phthalates," European Chemicals Agency, accessed February 8, 2025, https://echa.europa.eu/hot-topics/phthalates#:~:text=Phthalatesthat areclassifiedas,ofarticlescontainingthesephthalates.

152 **"more myth than reality"**: Olatz Fínex Marañon et al., *From Risk to Resilience*.

152 **"exposed to alarmingly" . . . adolescents tested**: Greet Schoeters, "All Europeans Are Exposed to Chemical Substances," VITO, May 2, 2022, https://vito.be/en/news/all-europeans-are-exposed-chemical-substances.

153 **hailed as a "great detox"**: European Environmental Bureau, "The Great Detox."

153 **"a lot of political backtracking"**: Vicky Cann, campaigner at Corporate Europe Observatory, email to author, January 21, 2025.

153 **"cut-price system"**: Roz Bulleid, Green Alliance, as quoted in Helena Horton, "Toxic Chemicals Banned by EU Since Brexit Still in Use in UK," *The Guardian*, September 15,

NOTES

2023, https://www.theguardian.com/environment/2023/sep/15/toxic-chemicals-banned-by-eu-since-brexit-still-in-use-in-uk.

154 **In a 2020 poll . . . regulations on chemicals:** Unchecked UK, "Attitudes of Younger Leave Voters to Regulation and Deregulation," Ipsos MORI Poll for Unchecked UK, May 2020, https://www.unchecked.uk/wp-content/uploads/2020/05/Attitudes-of-Younger-Leave-Voters-to-Regulation-and-Deregulation.pdf.

154 **"not only will there be no . . . strengthen" protections:** Michael Gove, "Green Brexit: A New Era for Farming, Fishing and the Environment," UK Department for Environment, Food and Rural Affairs, March 15, 2018, https://www.gov.uk/government/speeches/green-brexit-a-new-era-for-farming-fishing-and-the-environment.

155 **"failure-by-design" . . . funders:** Monbiot, "Britain Is Becoming a Toxic Chemical Dumping Ground."

155 **forty thousand tires' worth:** Sharon Lerner, "Toxic PFAS Chemicals Found in Artificial Turf," *The Intercept*, October 8, 2019, https://theintercept.com/2019/10/08/pfas-chemicals-artificial-turf-soccer/.

156 **One study estimated . . . Bay of Bengal:** Janice Brahney et al., "Plastic Rain in Protected Areas of the United States," *Science* 368, no. 6496 (June 2020): 1257–60, https://doi.org/10.1126/science.aaz5819; Dušan Materić et al., "Nanoplastics Transport to the Remote, High-Altitude Alps," *Environmental Pollution* 288 (November 2021): 117697, https://doi.org/10.1016/j.envpol.2021.117697; Rachel Hurley, Jamie Woodward, and James J. Rothwell, "Microplastic Contamination of River Beds Significantly Reduced by Catchment-Wide Flooding," *Nature Geoscience* 11 (March 2018): 251–57, https://doi.org/10.1038/s41561-018-0080-1; Imogen E. Napper et al., "The Abundance and Characteristics of Microplastics in Surface Water in the Transboundary Ganges River," *Environmental Pollution* 274 (April 2021): 116348, https://doi.org/10.1016/j.envpol.2020.116348; and Matt Simon, *A Poison Like No Other: How Microplastics Corrupted Our Planet and Our Bodies* (Island Press, 2022), 80, 112.

156 **75 percent of samples:** Antonio Ragusa et al., "Raman Microspectroscopy Detection and Characterisation of Microplastics in Human Breastmilk," *Polymers* 14, no. 13 (June 2022): 2700, https://doi.org/10.3390/polym14132700.

156 **Stool analyses . . . higher for children:** Simon, *A Poison Like No Other*, 4.

157 **Analyzing plaque . . . 60 percent:** Raffaele Marfella et al., "Microplastics and Nanoplastics in Atheromas and Cardiovascular Events," *New England Journal of Medicine* 390, no. 10 (March 2024): 900–910, https://doi.org/10.1056/nejmoa2309822.

158 **seven grams of plastic—roughly as much as five bottle caps, or a disposable spoon:** Nina Agrawal, "What Are Microplastics Doing to Our Bodies? This Lab Is Racing to Find Out," *New York Times*, April 8, 2025, https://www.nytimes.com/2025/04/08/well/microplastics-health.html.

158 **"I never would have imagined" . . . "plastic in my brain":** Susanne Rust, "As Global Plastic Production Grows, So Does the Concentration of Microplastics in Our Brains," *Los Angeles Times*, February 3, 2025, https://www.latimes.com/environment/story/2025-02-03/microplastics-study-plastic-production-dementia.

158 **His team found . . . without the diagnosis:** Alexander J. Nihart et al., "Bioaccumulation of Microplastics in Decedent Human Brains," *Nature Medicine* 31 (February 2025): 1114–19, https://doi.org/10.1038/s41591-024-03453-1.

NOTES

159 "Microplastic isn't a monolith" . . . living room floor daily: Simon, *A Poison Like No Other*, 4–5.

159 six thousand plastic fragments annually just in salt: Evangelos Danopoulos et al., "Microplastic Contamination of Salt Intended for Human Consumption: A Systematic Review and Meta-Analysis," *SN Applied Sciences* 2 (November 2020), https://doi.org/10.1007/s42452-020-03749-0.

159 Someone who drinks only . . . half a million: Evangelos Danopoulos, Maureen Twiddy, and Jeanette M. Rotchell, "Microplastic Contamination of Drinking Water: A Systematic Review," *PLoS One* 15, no. 7 (July 2020), https://doi.org/10.1371/journal.pone.0236838.

160 average tire loses . . . lifetime: Damian Carrington, "Car Tyres Are a Major Source of Ocean Microplastics—Study," *The Guardian*, July 14, 2020, https://www.theguardian.com/environment/2020/jul/14/car-tyres-are-major-source-of-ocean-microplastics-study.

160 thirty-one giant . . . long: Pieter Jan Kole et al., "Wear and Tear of Tyres: A Stealthy Source of Microplastics in the Environment," *International Journal of Environmental Research and Public Health* 14, no. 10 (October 2017): 1265, https://doi.org/10.3390/ijerph14101265.

160 "pigment suspended in liquid plastic": Burgess Brown, "What Do Plastic and Paint Have in Common? Everything," *Architizer*, https://architizer.com/blog/practice/materials/microplastic-plastic-paint-alternatives/.

160 Upward of one hundred thousand . . . load of laundry: Simon, *A Poison Like No Other*, 33–34.

160 giving off climate-warming emissions themselves as they degrade: Liz Plascencia, "More Than an Eyesore: Plastic Pollution's Contribution to Global Greenhouse Gas Emission," *Yale Environment Review*, June 23, 2022, https://environment-review.yale.edu/more-eyesore-plastic-pollutions-contribution-global-greenhouse-gas-emission.

161 "This stuff is increasing in our world exponentially": Agrawal, "What Are Microplastics Doing?"

161 "I don't need . . . quadruple": Rust, "As Global Plastic Production Grows."

Chapter 7: Wilson vs. Formosa

164 dolphins and oysters . . . her own shrimp boat: Diane Wilson, *Diary of an Eco-Outlaw: An Unreasonable Woman Breaks the Law for Mother Earth* (Chelsea Green, 2011), 5–10.

164 worst in the nation for toxic chemicals disposed on land: Toxic Release Inventory Program Division, *How Are the Toxic Release Inventory Data Used?* (US Environmental Protection Agency, May 2003), 7, https://nepis.epa.gov/Exe/ZyNET.exe/900B0I00.TXT?ZyActionD=ZyDocument&Client=EPA&Index=2000+Thru+2005&Docs=&Query=&Time=&EndTime=&SearchMethod=1&TocRestrict=n&Toc=&TocEntry=&QField=&QFieldYear=&QFieldMonth=&QFieldDay=&IntQFieldOp=0&ExtQFieldOp=0&XmlQuery=&File=D%3A%5Czyfiles%5CIndex%20Data%5C00thru05%5CTxt%5C00000011%5C900B0I00.txt&User=ANONYMOUS&Password=anonymous&SortMethod=h%7C-&MaximumDocuments=1&FuzzyDegree=0&ImageQuality=r75g8/r75g8/x150y150g16/i425&Display=hpfr&DefSeekPage=x&SearchBack=ZyActionL&Back=ZyActionS&BackDesc=Results%20page&MaximumPages=1&ZyEntry=1&SeekPage=x&ZyPURL.

NOTES

165 "Ms. Wilson, are you aware of this?": Diane Wilson, *An Unreasonable Woman: A True Story of Shrimpers, Politicos, Polluters, and the Fight for Seadrift, Texas* (Chelsea Green, 2005), 63.

166 too poor to buy him shoes: Jean-François Tremblay, "Formosa Plastics: Legendary Founder and Chairman of Taiwan's Formosa Plastics Group Steps Down," *Chemical and Engineering News*, June 12, 2006, https://cen.acs.org/articles/84/i24/Formosa-Plastics.html.

166 "a symbol of the island's economic miracle": Kathrin Hille, "Formosa Plastics Tycoon Hands Over to Children," *Financial Times*, June 5, 2006, https://www.ft.com/content/56b4d048-f48d-11da-86f6-0000779e2340.

167 "It was like riding a tiger" . . . "or get eaten": Louis Kraar, "The Overseas Chinese They Love the Getting, Not the Spending," *Fortune*, October 12, 1987, https://money.cnn.com/magazines/fortune/fortune_archive/1987/10/12/69646/index.htm.

167 He founded another company . . . backing for its imitation leather: Morten Bennedsen et al., *The Legacy of YC Wang* (INSEAD Publishing, 2019).

167 the brothers built . . . and gasoline: "Formosa Plastics Group Introduction," 2018, Formosa Plastics Group, https://www.fpg.com.tw/uploads/images/media-center/ebook-top/FPG%20Introduction2018_en.pdf.

167 His dark blue suits were perfectly ironed . . . 3:30 a.m. to exercise: Alice Yang, "Wang Yung-Ching: A Life Ends, the Legend Lives On," trans. Susanne Ganz, *CommonWealth Magazine*, October 23, 2008, https://english.cw.com.tw/article/article.action?id=1276.

167 his wife had to sneak new shirts . . . resisted spending money on clothes: Kraar, "The Overseas Chinese."

167 "Wang was harsh . . . their subordinates": Yang, "Wang Yung-Ching."

168 Taiwan's second-richest person, worth about $6.8 billion: "Taiwan's 40 Richest," *Forbes*, June 6, 2008, https://www.forbes.com/global/2008/0616/037.html?sh=1fe7627921db.

168 world's tenth-largest chemical company: Jean-François Tremblay, "Wang Yung-Ching, Founder of Formosa Plastics, Dies," *Chemical and Engineering News*, October 21, 2008, https://cen.acs.org/articles/86/web/2008/10/Wang-Yung-Ching-Founder-Formosa.html.

168 ninth biggest, with $31 billion in annual chemical sales: Alexander H. Tullo, "C&EN's Global Top 50 Chemical Firms for 2024," *Chemical and Engineering News,* July 19, 2024, https://cen.acs.org/business/finance/CENs-Global-Top-50-2024/102/i22.

168 Evidence emerged . . . taxes in Taiwan: CommonWealth Staff, "The Undiscovered Asian Offshore Tax Haven," *CommonWealth Magazine*, May 8, 2018, https://english.cw.com.tw/article/article.action?id=1939.

168 It filed a criminal defamation complaint . . . criticism of powerful corporations: Michele Catanzaro, "Scientist Cleared of Libel in Taiwan Court," *Nature News,* September 4, 2013, https://www.nature.com/articles/nature.2013.13685.

168 "The structure of Formosa Plastics Group" . . . "violations of regulations": Jane Patton et al., *Formosa Plastics Group: A Serial Offender of Environmental and Human Rights* (Center for International Environmental Law, the Center for Biological Diversity and Earthworks, October 2021), https://www.ciel.org/wp-content/uploads/2021/10/Formosa-Plastics-Group_A-Serial-Offender-of-Environmental-and-Human-Rights.pdf.

169 At home in Taiwan, rapid industrialization . . . as the project was announced: Peter Elkind, "The Wooing of Chairman Wang," *Texas Monthly*, February 1989, https://www.texasmonthly.com/news-politics/the-wooing-of-chairman-wang/.

NOTES

170 **fourth major expansion . . . had become twenty:** "Our Operations," Formosa Plastics Corporation, Texas, accessed December 6, 2023, https://www.fpctx.com/component/content/article?id=11&Itemid=181; and "Our Operations," Formosa Plastics Corporation USA, accessed December 6, 2023, https://www.fpg.taipei/en/issue-news/2022/340.

170 **about 2,500 Formosa employees and 1,000 contract hires:** Amy Blanchett, Formosa spokeswoman, email to author, January 24, 2022.

170 **enough energy to heat and power more than four hundred thousand homes:** Jim Vallette et al., *The New Coal: Plastics and Climate Change* (Beyond Plastics at Bennington College, October 2021), https://static1.squarespace.com/static/5eda91260bbb7e7a4bf528d8/t/616ef29221985319611a64e0/1634661022294/REPORT_The_New-Coal_Plastics_and_Climate-Change_10-21-2021.pdf.

170 **"basically anything that you can think of that's plastic":** Testimony of Richard Crabtree, San Antonio Bay Estuarine Waterkeeper and Sylvia Diane Wilson vs. Formosa Plastics Corp., Texas, and Formosa Plastics Corp., USA, in the US District Court Southern District of Texas, Victoria Division, trial transcript, vol. 4, 8.

171 **pledged fifty dollars . . . "Corporate greed kills":** Wilson, *Diary of an Eco-Outlaw*, 114–15.

173 **says it is committed to protecting the environment and guarding workers' health and safety:** "Our Commitment," Formosa Plastics Corporation, Texas, accessed March 21, 2023, https://www.fpctx.com/sustainability/commitment-to-ehs.

174 **an explosion at the plant . . . turning a small accident into a major disaster:** "CSB Issues Case Study of Formosa Plastics Point Comfort, Texas, Fire and Explosions," US Chemical Safety Board, July 20, 2006, https://www.csb.gov/csb-issues-case-study-of-formosa-plastics-point-comfort-texas-fire-and-explosions-unprotected-piping-non-fireproofed-structures-lack-of-automatic-shutoff-valves-noted-as-causes-flame-resistant-clothing-recommended/.

174 **in 2009 the company agreed . . . facility in Baton Rouge, Louisiana:** "Formosa Plastics Agrees to Resolve Multiple Environmental Violations at Plants in Texas and Louisiana," US Environmental Protection Agency, September 29, 2009, https://www.epa.gov/archive/epapages/newsroom_archive/newsreleases/6aef3a2a3a324b11852576410053a31c.html.

174 **As recently as 2021 . . . third-degree burns:** "Texas Plastics Corporation Will Pay Nearly $3 Million for Violating Clean Air Act," US Department of Justice, October 4, 2021, https://www.justice.gov/opa/pr/texas-plastics-corporation-will-pay-nearly-3-million-violating-clean-air-act.

175 **To end the spills . . . prove just how egregious they were:** Amy Johnson, consultant to Texas RioGrande Legal Aid and lawyer to Diane Wilson, interview with author, January 17, 2022.

176 **"We started wading" . . . "not a trace amount":** As quoted in Beth Gardiner, "How a Dramatic Win in Plastic Waste Case May Curb Ocean Pollution," *National Geographic*, February 23, 2022, https://www.nationalgeographic.com/environment/article/how-a-dramatic-win-in-plastic-waste-case-may-curb-ocean-pollution.

177 **Cheyenne said he'd found "handfuls" . . . "just dumped it":** Deposition testimony of Cheyenne Jurasek, "Plaintiff's Proposed Findings of Fact," filed March 18, 2019, San Antonio Bay Estuarine Waterkeeper and Wilson vs. Formosa Plastics Corp., 11, https://

NOTES

staticl.squarespace.com/static/5b58f65a96d455e767cf70d4/t/5c914b9f53450a91e7a5e708/1553025954088/Plaintiffs%27_Proposed_Findings_of_Fact.pdf.

178 **"demonstrate that plastic" ... in 2018:** Jeremy L. Conkle for Texas RioGrande Legal Aid, "Plaintiffs' Trial Exhibit No. 33," San Antonio Bay Estuarine Waterkeeper and Wilson vs. Formosa Plastics Corp., July 9, 2018, 5–10, 31, https://www.tceq.texas.gov/downloads/permitting/water-quality-standards-implementation/conkle-dr-ex-33.pdf; and Conkle for Texas RioGrande Legal Aid, "Addendum 2," San Antonio Bay Estuarine Waterkeeper and Wilson vs. Formosa Plastics Corp., March 4, 2019, 2, https://staticl.squarespace.com/static/5b58f65a96d455e767cf70d4/t/5c95005feb393122e02734c9/1553268839554/Conkle.Addendum.3.19.pdf.

178 **A manager from ... six other times:** "Plaintiff's Proposed Findings of Fact," San Antonio Bay Estuarine Waterkeeper and Wilson vs. Formosa Plastics Corp., 31, 39, 53, 75, 78; testimony of Stephen Ravel, trial transcript, vol. 1, 32; testimony of John Hyak, trial transcript, vol. 3, 127; testimony of Van Rozner, trial transcript, vol. 2, 127; and Rozner's photos from trial exhibit 259, provided to author by Johnson.

180 **"It would literally take me all day to describe to you all the improvements that Formosa made":** Opening statement of Diana Nichols, San Antonio Bay Estuarine Waterkeeper and Wilson vs. Formosa Plastics Corp., trial transcript, vol. 1, 37.

181 **All told, the judge determined ... "Formosa is a serial offender":** "Memorandum and Order," San Antonio Bay Estuarine Waterkeeper et al., vs. Formosa Plastics Corp., Texas, et al., US District Court Southern District of Texas, Victoria Division, June 27, 2019, https://staticl.squarespace.com/static/5b58f65a96d455e767cf70d4/t/5d1a76c3a0cd2a000115d463/1562015428807/liability+ruling+%282%29.pdf.

182 **largest-ever settlement of a private citizen's lawsuit under the Clean Water Act:** Johnson, citing research by Wilson's legal team, email to author, January 20, 2022.

182 **more than $25 million in penalties:** "Post-Consent Decree Violations and Environmental Mitigation Payment Records to the San Antonio Bay Estuarine Trust," provided to author by Diane Wilson via email, February 3, 2025.

182 **"The reverberations have been far-reaching":** As quoted in Gardiner, "Dramatic Win."

183 **$29 million for just the first phase of nurdle removal:** DD Turner, "Cox Creek Cleanup to Begin," *Port Lavaca Wave*, September 21, 2022, http://www.portlavacawave.com/news/around_town/cox-creek-cleanup-to-begin/article_0dc02000-395d-11ed-9a90-af327164f027.html.

183 **"No company wants the liability that Formosa Plastics found itself with":** As quoted in Gardiner, "Dramatic Win."

183 **More than seven hundred million:** Tristan Baurick, "A New Crop of Nurdles Is Washing Up in New Orleans, Renewing Calls for Cleanup, Penalties," *The Times-Picayune / New Orleans Advocate*, February 1, 2021, https://www.nola.com/news/environment/article_820710fc-64cc-11eb-85ce-93b2793068d5.html.

183 **Forty-nine tons' worth ... in Durban, South Africa:** "Update on Plastic Nurdles Incident Along Kwazulu Natal Coastline," Department of Environmental Affairs, South Africa, October 27, 2017, https://www.gov.za/news/media-statements/environmental-affairs-plastic-nurdles-incident-along-kwazulu-natal-coastline.

183 **some 1,700 metric tons' worth ... turtles and whales:** Katherine Bourzac, "Grappling with the Biggest Marine Plastic Spill in History," *Chemical and Engineering News*, January

NOTES

22, 2023, https://cen.acs.org/environment/pollution/marine-plastic-spill-xpress-pearl-nurdle/101/i3; and Zinara Rathnayake, "'Nurdles Are Everywhere': How Plastic Pellets Ravaged a Sri Lankan Paradise," *The Guardian*, January 25, 2022, https://www.theguardian.com/environment/2022/jan/25/nurdles-are-everywhere-how-plastic-pellets-ravaged-a-sri-lankan-paradise.

183 **British plants lose between five billion and fifty-three billion**: George Cole and Chris Sherrington, "Study to Quantify Plastic Pellet Loss in the UK," Eunomia Research and Consulting, February 11, 2016, https://eunomia.eco/reports/study-to-quantify-pellet-emissions-in-the-uk/.

Chapter 8: "They Want to Crush This Bug"

187 **"They Want to Crush This Bug"**: Josh Barbanel, "Suffolk County's Ban on Plastics Loses Allies," *New York Times*, December 31, 1991, https://www.nytimes.com/1991/12/31/nyregion/suffolk-county-s-ban-on-plastics-loses-allies.html.

187 **Sten Gustaf Thulin always . . . "why wouldn't you?"**: Laura Foster, "Plastic Carrier Bags: Why They Were Meant to Save the Planet," *BBC News*, October 17, 2019, https://www.bbc.co.uk/news/av/science-environment-50043369.

187 **The invention . . . serve as handles**: Sarah Laskow, "How the Plastic Bag Became So Popular," *The Atlantic*, October 10, 2014, https://www.theatlantic.com/technology/archive/2014/10/how-the-plastic-bag-became-so-popular/381065/.

188 **one of the most ubiquitous . . . "a thousand times its weight"**: Susan Freinkel, *Plastic: A Toxic Love Story* (Houghton Mifflin Harcourt, 2011), 143–45.

188 **a $600 million market . . . "We are going after that"**: Jube Shiver Jr., "Supermarket Dilemma: Battle of the Bags: Paper or Plastic?," *Los Angeles Times*, June 13, 1986, https://www.latimes.com/archives/la-xpm-1986-06-13-mn-10728-story.html.

188 **"a lot of re-educating to get people to accept plastic"**: Robert Siegel, as quoted in Kirk Johnson, "Plastic Producers Make Bid for Traditional Paper Market," *New York Times*, April 19, 1982, https://www.nytimes.com/1982/04/19/business/plastic-producers-make-bid-for-traditional-paper-market.html.

188 **industry set up the Plastic Grocery Sack Council . . . pack items in the new bags**: Freinkel, *Plastic*, 144.

188 **When Mobil's salesmen . . . still intact**: Martha M. Hamilton, "Resistance Gradually Melting: Future of Plastic Sacks in the Bag," *Los Angeles Times*, March 13, 1988, https://www.latimes.com/archives/la-xpm-1988-03-13-fi-1790-story.html.

189 **"couldn't compete . . . so expensive"**: Shiver, "Supermarket Dilemma."

189 **"it was pretty" . . . "landscape"**: Peter Grande, of Command Packaging, as quoted in Freinkel, *Plastic*, 145.

189 **5 percent of the US bag market . . . "growing by double-digit percentages a year"**: Hamilton, "Resistance Gradually Melting"; and "Suffolk Votes a Bill to Ban Plastic Bags," *New York Times*, March 30, 1988, https://www.nytimes.com/1988/03/30/nyregion/suffolk-votes-a-bill-to-ban-plastic-bags.html.

189 **five hundred bags a year . . . every minute**: Kitt Doucette, "The Plastic Bag Wars," *Rolling Stone*, July 25, 2011, https://www.rollingstone.com/politics/politics-news/the-plastic-bag-wars-243547/.

NOTES

189 **five trillion plastic bags are used annually**: "Our Planet Is Choking on Plastic," United Nations Environment Programme, accessed January 18, 2024, https://www.unep.org/interactives/beat-plastic-pollution.

189 **"Our job was and should always be to open plastics markets and keep them open"**: Jerome Heckman, as quoted in Freinkel, *Plastic*, 143.

189 **"Does society really need plastic bags?"**: Barry Commoner, as quoted in Shiver, "Supermarket Dilemma."

190 **Mobil and Amoco said discarded plastic was harmless**: "Suffolk Votes a Bill," *New York Times*.

190 **reduced pollution . . . other types of garbage**: John Rather, "Suffolk Weighs Plastics Delay," *New York Times*, May 7, 1989, https://www.nytimes.com/1989/05/07/nyregion/suffolk-weighs-plastics-delay.html.

191 **"Maybe we won't . . . on a barge"**: Suffolk County Executive Patrick Halpin, as quoted in "Suffolk Votes a Bill," *New York Times*.

191 **"All this bill does is discriminate" . . . "any solutions"**: Allen Gray, Mobil Oil Corporation spokesman, as quoted in Philip S. Gutis, "Plastic Wrapping Banned in Suffolk," *New York Times*, April 30, 1988, https://www.nytimes.com/1988/04/30/nyregion/plastic-wrapping-banned-in-suffolk.html.

191 **first-ever ban on plastic bags and containers**: Kevin Sack, "Appeals Court Upholds Suffolk's Pioneering Plastics Ban," *New York Times*, May 10, 1991, https://www.nytimes.com/1991/05/10/nyregion/appeals-court-upholds-suffolk-s-pioneering-plastics-ban.html.

191 **"They want to crush this bug, right here where it first started"**: Barbanel, "Suffolk County's Ban."

191 **"Several years . . . solid waste/packaging war"**: Tik Root, "Inside the Long War to Protect Plastic," Center for Public Integrity, May 16, 2019, https://publicintegrity.org/environment/pollution/pushing-plastic/inside-the-long-war-to-protect-plastic/.

191 **"I am going to lift up my plastic-foam cup and salute the judges"**: Leonard Raskin, of the Consumer Packaging Coalition, as quoted in Dennis Hevesi, "Ban on Plastics in Suffolk Is Overturned," *New York Times*, March 4, 1990, https://www.nytimes.com/1990/03/04/nyregion/ban-on-plastics-in-suffolk-is-overturned.html.

192 **"separates the plastic" . . . "markets are there and waiting"**: Rather, "Suffolk Weighs Plastics Delay."

192 **"The body is in the ground" . . . "on the headstone"**: Michael O'Donohoe, as quoted in Josh Barbanel, "Vote Blocks Plastics Ban for Suffolk," *New York Times*, March 4, 1992, https://www.nytimes.com/1992/03/04/nyregion/vote-blocks-plastics-ban-for-suffolk.html.

193 **$250 million ad campaign . . . "least afford it"**: Freinkel, *Plastic*, 149–52, 263.

194 **"Right before our eyes we see habits changing for the better"**: Ross Mirkarimi, as quoted in Stuart Glascock, "Seattle Looking to Dump Throwaway Bags," *Los Angeles Times*, April 14, 2008, https://www.latimes.com/archives/la-xpm-2008-apr-14-na-bags14-story.html.

194 **use soon quadrupled**: Freinkel, *Plastic*, 157.

194 **"were going to every" . . . "help kill it"**: Genevieve Abedon, of Ecoconsult, in Sacramento, California, and Tricia Cortez, executive director of the Rio Grande International

NOTES

Study Center, as quoted in James Osborne, "As Plastic Bans Spread, Industry Went on Attack," *Houston Chronicle*, August 3, 2019, https://www.houstonchronicle.com/business/energy/article/As-plastic-bans-spread-industry-went-on-attack-14273378.php.

194 **Chemistry Council helped bankroll . . . scrap the charge:** Claire Thompson, "Controversy Heats Up Over Seattle's Proposed Disposable Bag Fee," *Grist*, August 8, 2009, https://grist.org/culture/2009-08-07-bag-fee/.

195 **80 percent of residents . . . dropped by 60 percent:** "Skip the Bag, Save the River," District of Columbia Department of Energy and Environment, accessed January 4, 2025, https://doee.dc.gov/bags#:~:text=In%202009%2C%20DC%20Council%20passed,retail%20and%20food%2Dserving%20businesses.

195 **"been subject to a campaign of myths, misinformation and exaggeration":** Heather Caliendo, "Ban the Bags vs. Bag the Bans: The Fight Rages On," *Plastics Today*, February 5, 2013, https://www.plasticstoday.com/plastics-processing/ban-the-bags-vs-bag-the-bans-the-fight-rages-on.

195 **industry eventually sued, or threatened . . . state legislature:** Jennie R. Romer and Shanna Foley, "A Wolf in Sheep's Clothing: The Plastics Industry's 'Public Interest' Role in Legislation and Litigation of Plastic Bag Laws in California," *Golden Gate University Environmental Law Journal* 5, no. 2 (May 2021): 377–437; and Nick Welsh, "Can Plastic Bags Save the Planet?," *Salon*, September 13, 2012, https://www.salon.com/2012/09/13/can_plastic_bags_save_the_planet/.

195 **"the Patron Saint of Plastic Bags":** Belinda Luscombe, "The Patron Saint of Plastic Bags," *Time*, July 27, 2008, https://time.com/archive/6913216/the-patron-saint-of-plastic-bags/.

195 **150 Californian municipalities . . . required fees for paper:** "Plastic Bags: Local Ordinances," Californians Against Waste, accessed January 4, 2025, https://www.cawrecycles.org/list-of-local-bag-bans.

195 **plowed $6 million into California lobbying in 2007 and '08:** Freinkel, *Plastic*, 160.

195 **Chemistry Council and Hilex Poly . . . for the period:** Susan Ferriss, "Battle Over Bag Bill Draws Big Bucks," *Sacramento Bee*, November 13, 2010, https://www.sanluisobispo.com/news/local/article39137640.html#storylink=cpy.

195 **"California is in trouble" . . . "never witnessed this kind of opposition":** Julia Brownley, as quoted in Timothy Sandoval, "Plastic Bag Ban: Chemical Industry Targets Lawmakers," *California Watch* and *HuffPost*, August 27, 2010, https://www.huffpost.com/entry/platic-bag-ban-chemical-i_n_697159.

196 **price tag $3 million . . . "they make money":** Mark Murray of Californians Against Waste, as quoted in Phil Matier and Andrew Ross, "Plastic Bag Industry Profits as It Faces Tough Battle over Ban," *San Francisco Chronicle*, March 2, 2015, https://www.sfchronicle.com/bayarea/matier-ross/article/Plastic-bag-industry-profits-as-it-faces-tough-6109021.php#/0.

197 **"We can't visit every little town in the United States":** Mark Daniels, American Progressive Bag Alliance chairman, as quoted in "Plastic Bag Restrictions Gaining Momentum Despite Objections by Plastics Industry," *Bloomberg Law*, March 29, 2012, https://news.bloomberglaw.com/environment-and-energy/plastic-bag-restrictions-gaining-momentum-despite-objections-by-plastics-industry.

199 **"We didn't want to have" . . . "more business-friendly":** Rachel Leingang, "Don't Tread on Us: Cities Complain the State Is Usurping Their Authority," *Arizona Capitol Times*,

NOTES

April 30, 2015, https://azcapitoltimes.com/news/2015/04/30/dont-tread-on-us-cities-complain-the-state-is-usurping-their-authority/.

199 **"the absolute epitome of government run amok"**: State Senator Steve Smith, as quoted in Matthew Hendley, "Bag Bans Outlawed in Arizona," *Phoenix New Times*, April 14, 2015, https://www.phoenixnewtimes.com/news/bag-bans-outlawed-in-arizona-7287278#:~:text=Before%20a%20final%20vote%20on,left%20up%20to%20local%20authorities.

199 **"an ideology of collectivism"**: Then-state Representative Warren Petersen, as quoted in Howard Fischer, "Court Halts Lawsuit over Plastic Bag Ban," *Arizona Capitol Times*, October 14, 2016, https://azcapitoltimes.com/news/2016/10/14/court-halts-lawsuit-over-plastic-bag-ban/.

199 **"The one thing that I've found" . . . "monetary type of notification"**: State Senate president Andy Biggs, as quoted in Howard Fischer, "Senate Government Committee Votes to Force Cities to Obey State Laws," *Arizona Capitol Times*, February 18, 2016, https://azcapitoltimes.com/news/2016/02/18/senate-government-committee-votes-to-force-cities-to-obey-state-laws/.

202 **fifty million of them a year**: Corey Woods, Joel Navarro, and Lauren Kuby, "State Legislature Shouldn't Make Decisions for Tempe," *Arizona Republic*, April 5, 2015, https://www.azcentral.com/story/opinion/op-ed/2015/04/05/veto-bill-let-tempe-make-decisions-tempe/25277529/.

202 **The judge acknowledged . . . "damned if they don't"**: Maricopa County Superior Court Judge Douglas Gerlach, as quoted in Fischer, "Court Halts Lawsuit."

203 **stopped localities . . . guarding LGBTQ rights**: Sophie Quinton, "Expect More Conflict Between Cities and States," *Stateline*, January 25, 2017, https://stateline.org/2017/01/25/expect-more-conflict-between-cities-and-states/.

203 **"corporations and far-right groups go to buy government policy"**: Arn Pearson and David Armiak, "After 50 Years, This Right-Wing Law Factory Is Crazier Than Ever," *American Prospect*, October 4, 2023, https://prospect.org/power/2023-10-04-alec-50-years-right-wing-law-factory/.

204 **"mission to remake . . . a time"**: "United States of ALEC," *Moyers & Company*, September 28, 2012, https://billmoyers.com/episode/united-states-of-alec/.

204 **abortion rights . . . Equal Rights Amendment**: Brian Duignan, "American Legislative Exchange Council," *Encyclopedia Britannica*, December 17, 2024, https://www.britannica.com/topic/American-Legislative-Exchange-Council.

204 **began working not long after with the Gun Owners of America**: T. R. Reid, "Conservatives Devise Strategy to Kill D.C. Amendment," *Washington Post*, February 10, 1979, https://www.washingtonpost.com/archive/local/1979/02/11/conservatives-devise-strategy-to-kill-dc-amendment/52913ac1-032c-47ab-925a-b2b5281d3e16/.

204 **ALEC's treasurer in the 1980s**: Brendan Fischer, "Not Just the NRA: Former ALEC Leader, the Head of Gun Owners of America, Sides with Shooter of Trayvon Martin," *PRWatch*, Center for Media and Democracy, April 9, 2012, https://www.prwatch.org/news/2012/04/11394/not-just-nra-former-alec-leader-head-gun-owners-america-sides-shooter-trayvon-mar.

204 **far-right militias, white supremacists, and neo-Nazis**: Alexander Zaitchik, "The Zealot Larry Pratt Is the Gun Lobby's Secret Weapon," *Rolling Stone*, July 14, 2014, https://www

NOTES

.rollingstone.com/culture/culture-news/the-zealot-larry-pratt-is-the-gun-lobbys-secret-weapon-87059/.

204 **briefing with President Ronald Reagan:** Angie Walker, memorandum, December 10, 1981, Tobacco Institute Records and Council for Tobacco Research Records, Master Settlement Agreement, https://www.industrydocuments.ucsf.edu/docs/gzjb0033.

204 **"ALEC must begin" . . . "private sector supporters":** Pearson and Armiak, "After 50 Years"; and "Meeting the Challenge, Ideas + Action = Results," November 15, 1996, R. J. Reynolds Records, Master Settlement Agreement, https://www.industrydocuments.ucsf.edu/tobacco/docs/#id=slgp0088.

204 **tens of thousands of dollars for ALEC membership:** Yvonne Wingett Sanchez and Rob O'Dell, "What Is ALEC? 'The Most Effective Organization' for Conservatives, Says Newt Gingrich," *USA Today*, April 5, 2019, https://www.usatoday.com/story/news/investigations/2019/04/03/alec-american-legislative-exchange-council-model-bills-republican-conservative-devos-gingrich/3162357002/.

204 **ExxonMobil, Shell, Coca-Cola . . . Walmart:** "United States of ALEC," *Moyers & Company*.

204 **one hundred dollars a year:** "Legislative Membership," American Legislative Exchange Council, accessed February 4, 2025, https://alec.org/membership-type/legislative-membership/.

204 **golf, NRA-sponsored shooting:** Pearson and Armiak, "After 50 Years."

204 **Cigar parties . . . R. J. Reynolds:** "United States of ALEC," *Moyers & Company*.

204 **stogies proffered on silver platters:** Molly Redden, "Meet Mark Pocan, the Original ALEC Spy," *New Republic*, July 25, 2012, https://newrepublic.com/article/105308/meet-mark-pocan-original-alec-spy.

205 **Each is chaired jointly . . . They vote as equals:** Sanchez and O'Dell, "What Is ALEC?"

205 **"actually co-authored" . . . "craft public policy":** Christopher Leonard, *Kochland: The Secret History of Koch Industries and Corporate Power in America* (Simon & Schuster, 2019), 274.

205 **two main themes . . . "profit-driven legislation":** "United States of ALEC," *Moyers & Company*.

205 **A Missouri state representative . . . *Arizona Republic* reported:** Sanchez and O'Dell, "What Is ALEC?"

205 **bail industry representative . . . "making a dollar":** "United States of ALEC," *Moyers & Company*.

206 **Bail bondsmen still hold a seat on ALEC's board:** "Leadership," American Legislative Exchange Council, accessed February 4, 2025, https://alec.org/about/leadership/.

206 **"It's a situation where" . . . Food and Drug Administration:** Utility lobbyist Tim Kichlane, as quoted in Leonard, *Kochland*, 275, citing the *Austin American-Statesman*; and Leonard, *Kochland*, 114.

206 **"Like ideological" . . . Media and Democracy:** Lisa Graves, "ALEC Exposed: The Koch Connection," *The Nation*, July 12, 2011, https://www.thenation.com/article/archive/alec-exposed-koch-connection/.

206 **poured millions . . . advisory council:** Connor Gibson and Lisa Graves, "Koch Industries and ALEC: A History of Documents," *Koch Docs*, May 21, 2021, https://kochdocs

NOTES

.org/2019/09/03/koch-industries-and-alec-a-history-of-documents/; and American Legislative Exchange Council, "Leadership."

207 **unravel state . . . solar panels:** Editorial Board, "The Koch Attack on Solar Energy," *New York Times*, April 26, 2014, https://www.nytimes.com/2014/04/27/opinion/sunday/the-koch-attack-on-solar-energy.html; and Mike De Souza, "An Inside Look at U.S. Think Tank's Plans to Undo Environmental Legislation," *Toronto Star*, August 24, 2014, https://www.thestar.com/news/canada/an-inside-look-at-u-s-think-tanks-plans-to-undo-environmental-legislation/article_aa08f625-b3fe-5f93-8caa-c34225115551.html.

207 **"to push back against woke" . . . gas, and coal:** Kate Aronoff, "Conservatives Have a New Bogeyman: Critical Energy Theory," *New Republic*, December 7, 2021, https://newrepublic.com/article/164641/conservatives-new-bogeyman-critical-energy-theory.

207 **"ALEC is a big reason . . . climate change":** Nick Surgey, of the Center for Media and Democracy, as quoted in De Souza, "An Inside Look."

207 **"Warming Up to Climate Change":** "United States of ALEC," *Moyers & Company*.

207 **casting doubt on established science:** Neela Banerjee, "What's Behind ALEC's Denial That It Denies Climate Change?," *Inside Climate News*, April 14, 2015, https://insideclimatenews.org/news/14042015/whats-behind-alecs-denial-it-denies-climate-change/.

207 **ALEC's climate views . . . to leave:** "Exxon Mobil Joins Exodus of Firms from Lobbying Group ALEC," Reuters, July 12, 2018, https://www.reuters.com/article/us-exxon-mobil-alec/exxon-mobil-joins-exodus-of-firms-from-lobbying-group-alec-idUSKBN1K231R/.

207 **Google quit . . . "literally lying":** Brian Fung, "Google: We're Parting with the Climate Change Skeptics at ALEC," *Washington Post*, September 22, 2014, https://www.washingtonpost.com/news/the-switch/wp/2014/09/22/google-were-parting-with-the-climate-change-skeptics-at-alec/.

207 **NPR reported . . . profit from detaining immigrants:** Laura Sullivan, "Shaping State Laws with Little Scrutiny," NPR, October 29, 2010, https://www.npr.org/transcripts/130891396.

208 **General Motors . . . with ALEC:** "United States of ALEC," *Moyers & Company*.

208 **still represented . . . Airbnb:** American Legislative Exchange Council, "Leadership"; and "About Us," NetChoice, accessed February 4, 2025, https://netchoice.org/about/#our-mission.

208 **nearly a quarter of all state lawmakers:** "About ALEC," American Legislative Exchange Council, accessed February 4, 2025, https://alec.org/about/.

208 **overwhelmingly Republican:** "ALEC Politicians," *SourceWatch*, September 19, 2019, https://www.sourcewatch.org/index.php/ALEC_Politicians.

208 **five members . . . House Speaker:** "Alumni," American Legislative Exchange Council, accessed May 23, 2025, https://alec.org/about/alumni/.

208 **In one eight-year period . . . six hundred became law:** Sanchez and O'Dell, "What Is ALEC?"

208 **At a 2014 meeting . . . boxes, cups, and bottles:** Steve Arnold, "Undercover at ACCE: ALEC Offshoot Spins City and County Officials on Dirty Energy, Local Control," *PRWatch*, Center for Media and Democracy, February 23, 2015, https://www.prwatch.org/news/2015/02/12747/undercover-acce-alecs-offshoot-spins-city-and-county-officials-dirty

NOTES

-energy; and Calvin Sloan and Jessica Mason, "Barring Plastic Bag Bans, Another ALEC Law Takes Aim at Local Democracy," *Exposed by CMD*, Center for Media and Democracy, March 16, 2016, https://www.exposedbycmd.org/2016/03/16/barring-plastic-bag-bans-another-alec-law-takes-aim-at-local-democracy/.

208 **"[Insert Jurisdiction]" ... "the container":** "Regulating Containers to Protect Business and Consumer Choice," American Legislative Exchange Council, accessed February 4, 2025, https://alec.org/model-policy/regulating-containers-to-protect-business-and-consumer-choice/.

208 **Before long, ALEC's board ... and other containers:** "Act to Establish Statewide Uniformity for Auxiliary Container Regulations," American Legislative Exchange Council, accessed February 4, 2025, https://alec.org/model-policy/act-to-establish-statewide-uniformity-for-auxiliary-container-regulations/; and Senate Bill 1241, State of Arizona Senate, Fifty-Second Legislature, First Regular Session, 2015, https://www.azleg.gov/legtext/52leg/1r/laws/0271.pdf.

209 **"It was not very good democracy":** Cam Gordon, as quoted in Scott Rodd, "Banning the Bans: State and Local Officials Clash over Plastic Bags," *Stateline*, January 29, 2018, https://stateline.org/2018/01/29/banning-the-bans-state-and-local-officials-clash-over-plastic-bags/.

209 **American Chemistry Council plowed $450,000 ... the law's passage:** "American Chemistry Council, Inc.," June 2017 Report, Minnesota Campaign Finance Board, https://cfb.mn.gov/reports-and-data/viewers/lobbying/lobbying-organizations/2750/2017.1/.

209 **American Progressive Bag Alliance chipped in $60,000:** "American Progressive Bag Alliance," June 2017 Report, Minnesota Campaign Finance Board, https://cfb.mn.gov/reports-and-data/viewers/lobbying/lobbying-organizations/7169/2017.1/.

209 **Phil Rozenski, an executive ... a 1993 Texas waste law:** Osborne, "As Plastic Bans Spread."

210 **eleven other local measures:** Sarah Gibbens, "See the Complicated Landscape of Plastic Bans in the U.S.," *National Geographic*, August 15, 2019, https://www.nationalgeographic.com/environment/2019/08/map-shows-the-complicated-landscape-of-plastic-bans/.

210 **prohibited in seventeen states:** "State Plastic Bag Legislation," National Council of State Legislatures, February 8, 2021, https://www.ncsl.org/environment-and-natural-resources/state-plastic-bag-legislation.

210 **"Plastic Bags Have ... Winning":** Samantha Maldonado, Bruce Ritchie, and Debra Kahn, "Plastic Bags Have Lobbyists. They're Winning," *Politico*, January 20, 2020, https://www.politico.com/news/2020/01/20/plastic-bags-have-lobbyists-winning-100587.

210 **more than $7 billion ... to $34 billion:** Ismail Sutaria, "Plastic Bag Market Trends—Growth & Forecast 2023–2033," Future Market Insights, 2024.

210 **"the stigmatization ... fight like hell":** Caliendo, "Ban the Bags."

211 **twelve states ... now bar:** California, Colorado, Connecticut, Delaware, Hawaii, Maine, New Jersey, New York, Oregon, Rhode Island, Vermont, and Washington.

212 **two executives from packaging maker Novolex ... "baby formula and dog food":** Dustin Gardiner, "How Industry 'Environmental' Group Helped Foil California's Plastics Crackdown," *San Francisco Chronicle*, October 1, 2019, https://www.sfchronicle.com/politics/article/How-a-plastics-environmental-group-helped-14483578.php.

NOTES

212 **funded their own . . . confuse consumers:** Alice Delemare Tangpuori et al., *Talking Trash* (Changing Markets Foundation, September 2020), 8, https://talking-trash.com/wp-content/uploads/2020/09/TalkingTrash_4_US2.pdf.

Chapter 9: "No Silver Bullet"

213 **"No Silver Bullet":** Johan Aurik et al., *No Silver Bullet: Why a Mix of Solutions Will Achieve Circularity in Europe's Informal Eating Out (IEO) Sector* (Kearney, 2023), https://nosilverbullet.eu/wp-content/uploads/2023/05/No-silver-bullet–Why-a-mix-of-solutions-will-achieve-circularity-in-Europes-informal-eating-out-IEO-sector.pdf.

213 **50 percent of the plane . . . "Dream Machine":** Stanley Holmes, "Boeing's Plastic Dream Machine," *Bloomberg Businessweek*, June 19, 2005, https://www.bloomberg.com/news/articles/2005-06-19/boeings-plastic-dream-machine?embedded-checkout=true&leadSource=uverify%20wall.

213 **"We pushed through an open door, in a sense":** Joan Marc Simon, Zero Waste Europe founder, interview with author, March 21, 2024.

213 **billions fewer used each year:** "Decrease in Lightweight Plastic Bags Continued in 2021," Eurostat, November 7, 2023, https://ec.europa.eu/eurostat/web/products-eurostat-news/w/ddn-20231107-1.

214 **"this thing that pretty much every person on the street thought was a good idea":** Vicky Cann, campaigner at Corporate Europe Observatory, interview with author, March 27, 2024.

215 **lobbyists pressed . . . analyzing the emails:** "Picking Up the Plastics Trail: How Ireland Cooperated with the Plastics Industry," Corporate Europe Observatory, November 13, 2019, https://corporateeurope.org/en/2019/11/picking-plastics-trail-how-ireland-cooperated-plastics-industry.

219 **"to be wandering around with a whole box of containers":** Ioana Popescu, Rethink Plastic alliance coordinator, interview with author, March 18, 2024.

219 **draft bill . . . other dangerous chemicals:** "Proposal for a Regulation of the European Parliament and of the Council on Packaging and Packaging Waste," EUR-Lex, November 30, 2022, https://eur-lex.europa.eu/legal-content/EN/TXT/?uri=CELEX%3A52022PC0677.

221 **packaging makers Amhil and Novolex:** "Our Members," 360° Foodservice, accessed February 9, 2025, https://www.360foodservice.com/en/our-members/aug/.

221 **study titled *No Silver Bullet* . . . need for dishwashing:** Aurik et al., *No Silver Bullet*.

221 **"This well-meaning regulation" . . . water consumption:** "Major New Report Launch," No Silver Bullet, March 1, 2023, https://nosilverbullet.eu/wp-content/uploads/2023/03/Press-Release.EPPAFinal_03March.pdf.

221 **"The idea of reusing" . . . more than one material:** Jon Banner, "No Silver Bullet: Packaging Waste Plans Wrong to Prioritize Reuse over Recycling, Finds Report," *Politico*, September 28, 2023, https://www.politico.eu/sponsored-content/no-silver-bullet-packaging-waste-plans-wrong-to-prioritize-reuse-over-recycling-finds-report/.

222 **Another industry-commissioned report:** "New Life Cycle Analysis Shows That Recyclable, Paper-Based Packaging Used in the Quick Service Food Delivery and Takeaway Sector Offers Significant Environmental Benefits," European Paper Packaging Alliance, November 2022, https://eppa-eu.org/lca-study-on-takeaway/.

NOTES

222 **more than fifty experts . . . skewed their conclusions:** Alba Bala Gala et al., "Life Cycle Assessment Scientists Urge EU Policy Makers to Treat Some Packaging Environmental Impact Assessments with Caution," accessed February 9, 2025, https://www.politico.eu/wp-content/uploads/2023/09/19/open-letter-LCA-packaging97.pdf.

222 **"at best flawed, at worst deliberately misleading":** Dario Cottafava, "Misleading Packaging Studies: Open Letter from LCA Experts," *Medium*, September 18, 2023, https://medium.com/@dariocottafava/misleading-packaging-studies-open-letter-from-lca-experts-86365f47ec4f.

222 **official assessment . . . throwaway paper packaging:** "Questions & Answers on the Regulation on Packaging and Packaging Waste," European Commission, November 30, 2022, https://ec.europa.eu/commission/presscorner/detail/en/qanda_22_7157.

222 **"The best sustainability solutions" . . . "the recycling one":** No Silver Bullet, "Major New Report Launch."

223 **Seda is one . . . Huhtamaki:** Clare Carlile, "McDonald's Leads Lobbying Offensive Against Laws to Reduce Packaging Waste in Europe," *DeSmog*, May 8, 2023, https://www.desmog.com/2023/05/08/mcdonalds-leads-lobbying-offensive-against-laws-to-reduce-packaging-waste-in-europe/.

223 **"does not give in to solutions that penalize our industry":** Simone De La Feld, "EU Ambassadors' Go-Ahead for Agreement on Packaging Regulation," *EUNews*, March 15, 2024, https://www.eunews.it/en/2024/03/15/eu-ambassadors-go-ahead-for-agreement-on-packaging-regulation-italy-system-cheers/.

223 **They proposed . . . other sectors:** PPWR—plenary AMS, p. 200, Amendment 513; ENVI-AM-DA 1708 A 2007, p. 50, Amendment 1800; p. 53, Amendment 1805; and p. 63, Amendment 1829; ENVI-AM-da-2008 a 2307, p. 68, Amendment 2159; and communications from European Paper Packaging Alliance and Federazione Carta e Grafica, provided to author by a European Parliament source on condition of anonymity.

223 **eliminating the bans . . . L'Oréal, and Unilever:** PPWR—plenary AMS p. 80, Amendment 413; and "Plastics Packaging Value Chain Views on the Proposal for a Packaging & Packaging Waste Regulation," signed by A Circular Economy for Flexible Packaging (Ceflex), European Manufacturers of EPS, European Plastics Converters, Flexible Packaging Europe, Polyolefin Circular Economy Platform, Plastics Europe, Styrenics Circular Solutions, and Vinyl Plus, provided to author by a European Parliament source on condition of anonymity; and "Who We Are: Consortium Stakeholders," Ceflex, https://ceflex.eu/who-we-are/.

223 **Five other industry groups . . . respond to my queries:** European Organization for Packaging and Environment (EUROPEN), FoodDrinkEurope, Soft Drinks Europe (UNESDA), BusinessEurope, and Petrochemicals Europe.

224 **lobbyists had followed his colleagues into bathrooms:** Mohammed Chahim, as quoted in Leonie Cater, "Parliament Probing Lobbyists Who Fought Sustainable Packaging Rules," *Politico*, January 26, 2024, https://www.politico.eu/article/eu-parliament-qatargate-climate-lobbyists-sustainable-packaging-rules/.

224 **"The meat came off" . . . "outrageously bad"; "The way agreement . . . loopholes and exemptions":** Popescu, interview with author.

NOTES

Chapter 10: "An Impatient Billionaire"

229 "An Impatient Billionaire": Member of Flanders Parliament Mieke Schauvliege, as quoted in Lauren Walker, "'Threats Have Worked': Flanders Grants Permit for INEOS' Controversial Ethane Cracker After All," *Brussels Times*, January 8, 2024, https://www.brusselstimes.com/866644/threats-have-worked-flanders-grants-permit-for-ineos-controversial-ethane-cracker-after-all.

229 He's trekked to the North . . . six African countries: Dan Cancian, "Inside Sir Jim Ratcliffe's Super-Fit Lifestyle," *Daily Mail Online*, April 27, 2024, https://www.dailymail.co.uk/sport/football/article-13344719/Sir-Jim-Ratcliffe-Man-United-INEOS-London-Marathon.html.

230 broke his foot . . . "have not completed": Sylvia Pfeifer, "Jim Ratcliffe," *Financial Times*, November 20, 2014, https://www.ft.com/content/9318986c-8ec2-11e3-b6f1-00144feab7de.

230 "I used to think" . . . attend university; "I put all my chips"; "I'd have been in a mess": Jim Ratcliffe and Ursula Heath, *The Alchemists: The INEOS Story; An Industrial Giant Comes of Age* (Biteback Publishing, 2018) 54–55, 80.

230 more than doubled in value: Dearbail Jordan and Noor Nanji, "Jim Ratcliffe: Who Is the Man Buying Man Utd Stake?," *BBC News*, December 24, 2023, https://www.bbc.co.uk/news/business-44101223.

230 "Jim was eager" . . . "before adjusting": Pfeifer, "Jim Ratcliffe."

231 "We mopped up . . . bulk chemical businesses": Ratcliffe and Heath, *The Alchemists*, 10.

231 "His negotiating tactics . . . pain is inflicted": Pfeifer, "Jim Ratcliffe."

231 Innovene, a BP subsidiary . . . company lore: Ratcliffe and Heath, *The Alchemists*, 99–113.

232 "incredibly unstuffy and unbureaucratic": Pfeifer, "Jim Ratcliffe."

232 "You want to do" . . . "aggressive they are": Murad Ahmed and Michael Pooler, "Ineos: Why Jim Ratcliffe Is Mixing Petrochemicals and Sports," *Financial Times*, March 15, 2020, https://www.ft.com/content/31b5e7e0-5c76-11ea-b0ab-339c2307bcd4.

232 "I paid my taxes" . . . "I'm afraid": Tariq Panja, "A Billionaire Bought a Chunk of Manchester United. Now He Has to Fix It," *New York Times*, February 22, 2024, https://www.nytimes.com/2024/02/22/world/europe/manchester-united-jim-ratcliffe.html.

232 running at just 40 percent of capacity: Tom Crotty, director and spokesman at INEOS, interview with author, November 18, 2020.

233 more than tripled profits . . . in charge of the project; minus ninety degrees Celsius . . . two football fields: Ratcliffe and Heath, *The Alchemists*, 193–200, 206.

233 biggest labor dispute in decades . . . "large part of my community": Pfeifer, "Jim Ratcliffe."

234 produces a million and a half . . . polyethylene and polypropylene: "INEOS @ Grangemouth," INEOS, accessed September 6, 2024, https://www.ineos.com/sites/grangemouth/about/.

234 sixth-largest chemical company: Alexander H. Tullo, "C&EN's Global Top 50 Chemical Firms for 2023," *Chemical and Engineering News*, July 24, 2023, https://cen.acs.org/business/finance/CENs-Global-Top-50-2023/101/i24.

234 his net worth is estimated at more than £23 billion: "The Sunday Times Rich List 2024," *Sunday Times*, May 17, 2024, https://www.thetimes.com/sunday-times-rich-list.

NOTES

235 **Facebook and McDonald's combined**: Ratcliffe and Heath, *The Alchemists*, 3.

235 **"the biggest company you've never heard of"; "It's pretty special"... "significant sporting asset"**: Ahmed and Pooler, "Ineos."

235 **"when a private individual"... "questions to ask"**: Eoin Connolly, "At Large: Just What Are Sir Jim Ratcliffe's Plans in Sport?," *SportsPro*, May 9, 2019, https://www.sportspromedia.com/insights/opinions/sportspro-blog/team-ineos-kipchoge-sir-jim-ratcliffes-sport-sponsorship/.

236 **"throwing an awful lot of positive PR at the company"**: Graham Copley, C-MACC founding partner, as quoted in Ahmed and Pooler, "Ineos."

236 **"I would love"... "discussions with decision-makers"**: Delphine Lévi Alvarès, global petrochemicals campaign manager at the Center for International Environmental Law, interview with author, April 16, 2024.

236 **2019 BBC sports interview**: "Fracking Boss Jim Ratcliffe Hits Out at 'Pathetic' Government," *BBC News*, May 1, 2019, https://www.bbc.co.uk/news/business-48125581.

238 **$4 billion ... "possibly the world"**: "Project One," *INEOS Inch*, 2024, https://www.ineos.com/inch-magazine/articles/issue-25/project-one.

238 **Onstage at a 2023 ... "becomes carbon neutral"**: "Interview with Founder and Chairman Sir Jim Ratcliffe at the 'Vooruitblik 2023' Event by Voka—Kamer van Koophandel Antwerpen-Waasland," INEOS Project One, February 27, 2023, https://project-one.ineos.com/en/stories/interview-with-founder-and-chairman-sir-jim-ratcliffe-at-the-vooruitblik-2023-event-by-voka-kamer-van-koophandel-antwerpen-waasland/.

239 **"climate neutral" ... "Tear the place down again?"**: Walker, "'Threats Have Worked."

241 **Terrified INEOS would decide not to build in Antwerp**: "Nitrogen Crisis Deepens Rift in Flemish Government," Belga News Agency, August 1, 2023, https://www.belganewsagency.eu/nitrogen-crisis-deepens-rift-in-flemish-government.

242 **"The government wants" ... "no regard for us"**: Bart Dickens, Farmers Defense Force chairman, as quoted in *The Brussels Times* with Belga, "Farmers Protest in Antwerp Against Nitrogen Decree," *Brussels Times*, December 20, 2023, https://www.brusselstimes.com/belgium/846129/farmers-protest-in-antwerp-against-nitrogen-decree.

242 **"dancing to the tune of an impatient billionaire"**: Flemish member of Parliament Mieke Schauvliege, as quoted in Walker, "Threats Have Worked."

243 **"It was very hush-hush" ... "privileged access"**: Vicky Cann, campaigner at Corporate Europe Observatory, interview with author, March 27, 2024.

243 **"This is the last chance ... emissions"**: INEOS Group, "Sir Jim Ratcliffe: 'This Is the Last Chance to Stop Europe from Sleepwalking into Off-Shoring Its Industry, Jobs, Investments and Emissions,'" INEOS, February 20, 2024, https://www.ineos.com/news/ineos-group/sir-jim-ratcliffe—this-is-the-last-chance-to-stop-europe-from-sleepwalking-into-off-shoring-its-industry-jobs-investments-and-emissions/.

243 **in a communiqué they titled ... operating expenses**: "The Antwerp Declaration for a European Industrial Deal," https://antwerp-declaration.eu/.

245 **"What's really sad" ... "a big man now"**: Lévi Alvarès, interview with author.

245 **Ratcliffe visited ... called in to the meeting**: Walker, "Threats Have Worked."

245 **"If we had not seen" ... pulled the plug**: Ratcliffe interview with *De Tijd*, as quoted in Lisa O'Carroll and Sandra Laville, "Jim Ratcliffe's Vast Petrochemical Plant in Antwerp

NOTES

Faces New Legal Challenge," *The Guardian*, February 22, 2024, https://www.theguardian.com/environment/2024/feb/22/ngos-lodge-legal-challenge-against-vast-new-petrochemical-plant-in-antwerp.

245 "tsunami of goals" . . . "especially pleased": Statement of Annick De Ridder, Antwerp vice mayor and chairwoman of Port of Antwerp-Bruges Board of Directors, emailed to author, March 12, 2024.

245 Flanders's government gave . . . €700 million guarantee: "Ineos Secures €3.5 Billion Financing for Project One—the Greenest Cracker in Europe," INEOS, February 13, 2023, https://www.ineos.com/news/ineos-group/ineos-secures-3.5-billion-financing-for-project-one—-the-greenest-cracker-in-europe/.

245 "The project was deemed" . . . "management of these impacts": Sandra Laville, "UK Gives £600m Backing to Jim Ratcliffe's 'Carbon Bomb' Petrochemical Plant," *The Guardian*, February 29, 2024, https://www.theguardian.com/environment/2024/feb/29/uk-600m-backing-jim-ratcliffe-carbon-bomb-petrochemical-plant.

246 "It has a huge amount of momentum behind it": Cann, interview with author.

246 an INEOS video . . . ten days: "The Ethane Tank, the Logistical Heart of Project One," INEOS Project One, June 5, 2024, https://project-one.ineos.com/en/news/the-ethane-tank-the-logistical-heart-of-project-one/.

246 "bolted together like a big Lego set": Crotty, interview with author, September 19, 2024.

Chapter 11: Shipping Plastic, Shifting Blame

249 six times more: Mike Ives, "Recyclers Cringe as Southeast Asia Says It's Sick of the West's Trash," *New York Times*, June 7, 2019, https://www.nytimes.com/2019/06/07/world/asia/asia-trash.html.

249 jumped by 35 percent, even as their reported value rose by just 6 percent: Ecoton and the Party Department, *Take Back!*, January 24, 2020, video, https://www.youtube.com/watch?v=pP9JT_EE2NE.

250 she did not believe it was ever that high: Liana Bratasida, executive director of the Indonesian Pulp and Paper Association (APKI), interview with author, June 5, 2023.

251 A short documentary . . . "developed countries' landfill": Ecoton and the Party Department, *Take Back!*

252 "We were supposed to" . . . one such event, in 2019: Lucia Binding, "Indonesia Sends Back 100 Tonnes of Waste Including Plastics to US," *Sky News*, June 24, 2019, https://news.sky.com/story/indonesia-sends-back-100-tonnes-of-waste-including-plastics-to-us-11743446.

252 the United States, France, Spain, Greece, and Australia: Achmad Ibrahim and Niniek Karmini, "Indonesia Sending Back 547 Containers of Waste from West," Associated Press, September 18, 2019, https://apnews.com/general-news-42531a97a187493abdcce76bec338c90.

252 "I will advise Canada" . . . "Eat it if you want to": Jason Gutierrez, "Philippines Sets Deadline for Canada to Take Back Trash," *New York Times*, May 8, 2019, https://www.nytimes.com/2019/05/08/world/asia/duterte-philippines-canada-trash.html.

252 sent on to other poor nations, including India, Mexico, and Vietnam: "Global Waste Shell Game: 'Returned' Illegal Waste Shipments from U.S., Diverted from Indonesia to Other Asian Countries," Basel Action Network, October 28, 2019, https://www.ban.org

NOTES

/news-new/2019/10/28/global-waste-shell-game-returned-illegal-waste-shipments-from-us-diverted-from-indonesia-to-other-asian-countries.

252 **"only care about their pocket money"**: Arisman, Center for Southeast Asian Studies Indonesia executive director, interview with author, May 9, 2023, as quoted in Beth Gardiner, "Indonesia Cracks Down on the Scourge of Imported Plastic Waste," *Yale Environment 360*, August 1, 2023, https://e360.yale.edu/features/plastic-waste-imports-recycling-indonesia.

253 **backed down after . . . hard to achieve**: Corin Williams, "Indonesia to Allow Waste Imports with 2% Contamination Limit," *Materials Recycling World*, July 19, 2021, https://www.mrw.co.uk/news/indonesia-to-allow-waste-imports-with-2-contamination-limit-19-07-2021/.

253 **more plastic scrap from the Marshall Islands . . . United States**: Yuyun Ismawati Drwiega and Mochamad Adi Septiono, *Plastic Waste Trade in Indonesia Country Update Report* (Aliansi Zero Waste Indonesia, Ecoton, BaliFokus, Nexus3 Foundation, IPEN, Plastic Solutions Fund, and Break Free from Plastic), June 2019, https://16edd8c0-c66a-4b78-9ac3-e25b63f72d0f.filesusr.com/ugd/13eb5b_baf87bf54c7d4954ad706d0820ebdf7d.pdf?index=true.

254 **Activists believe it was waste . . . island chain**: Yuyun Ismawati Drwiega, interview with author, April 11, 2023; and Prigi Arisandi, Ecoton, interview with author, May 2, 2023.

254 **a mixed picture of halting progress . . . larger every year**: As quoted in Gardiner, "Indonesia Cracks Down."

257 **A woman in a dark pink blouse . . . bring her material to resell**: Daslah, daughter of the owner of HSR Putri Mandiri waste-sorting lot, Kragilan, Serang, Banten province, interview with author, May 6, 2023.

258 **"It's very exhausting"**: As quoted in Gardiner, "Indonesia Cracks Down."

258 **she tells me Indah Kiat's imports . . . burning its waste plastic as fuel**: Letchumi Achana, head of advocacy at Asia Pulp and Paper, Indah Kiat's parent company, Whats-App messages, July 3, 2023, as quoted in Gardiner, "Indonesia Cracks Down."

258 **comply with the 2 percent rule**: Bratasida, interview with author.

259 **"How come it's our problem?" . . . "help yourself"**: Ismawati Drwiega, interview, as quoted in Gardiner, "Indonesia Cracks Down."

261 **"are finding ways to wiggle out" . . . "we're not going to bother'"**: Gardiner, "Indonesia Cracks Down."

261 **shifted exports toward Latin America . . . began sending more to Turkey**: "Plastic Waste Trade Data," Basel Action Network, accessed December 8, 2023, https://www.ban.org/plastic-waste-transparency-project-hub/trade-data.

263 **single bar . . . "affordable to them"**: R. V. Rajan, "The Man Behind the Sachet Revolution in India," *Madras Musings*, https://www.madrasmusings.com/vol-30-no-17/the-man-behind-the-sachet-revolution-in-india/.

263 **in 1976 began using them to sell a shampoo he called Velvette**: Anand Kumar Jaiswal, "Cavinkare Private Limited: Serving Low Income Customers," *Asian Case Research Journal* 12, no. 1 (2008), https://ssrn.com/abstract=1603725.

263 **"This is going to be the product of the future"**: Rajan, "Man Behind the Sachet Revolution."

NOTES

263 **named his shampoo Chik . . . three-quarters of a rupee each:** PC Vinoj Kumar, "How a Small-Town Boy Who Lived a Carefree Life Built a Rs 1450 Crore Turnover Company," *Weekend Leader*, December 20, 2017, https://www.theweekendleader.com/Success/2685/the-giant-killer.html.

263 **He staged live demonstrations . . . free samples after the show:** Sobana Ravichandran, "The Inspiring Success Story of CavinKare," July 19, 2018, LinkedIn post, https://www.linkedin.com/pulse/inspiring-success-story-cavinkare-sobana-ravichandran/.

264 **"They came into the sachet business a decade after us":** Swati Anand and Rajesh Chandramauli, "CavinKare Builds Big on Sachet Branding," *Times of India*, December 4, 2010, http://timesofindia.indiatimes.com/articleshow/7039958.cms?utm_source=contentofinterest&utm_medium=text&utm_campaign=cppst.

264 **"We discovered that wealth lies in rural India" . . . nearly 70 percent of shampoo:** Joe Brock and John Geddie, "Unilever's Plastic Playbook," Reuters, June 22, 2022, https://www.reuters.com/investigates/special-report/global-plastic-unilever/.

264 **Filipinos, for example . . . one estimate:** Catherine Liamzon et al., "Sachet Economy: Big Problems in Small Packets," Global Alliance for Incinerator Alternatives, July 2020, https://www.no-burn.org/wp-content/uploads/2021/11/Sachet-Economy-spread-.pdf.

264 **just over a trillion . . . in a decade:** "Sachet Packaging Market: Global Industry Analysis 2018–2022 and Opportunity Assessment 2023–2033," Future Market Insights, 2023, 33.

264 **more shampoo is sold in sachets . . . other products too:** Ismail Sutaria, chief packaging consultant, Future Market Insights, interview with author, September 7, 2023.

264 **nearly $12 billion a year, and is expected to surpass $17 billion by 2033:** "Sachet Packaging Market," 33.

264 **"Sachet packaging has a bright future":** Ismail Sutaria, "Sachet Packaging Market Snapshot (2023 to 2033)," Future Market Insights, July 2021, https://www.futuremarketinsights.com/reports/sachet-packaging-market.

265 **"evil, because you cannot" . . . "'have no choice'":** Hanneke Faber, then-Unilever president for global food and refreshments, and Alan Jope, then-Unilever CEO, as quoted in Brock and Geddie, "Unilever's Plastic Playbook."

266 **only 65 percent of trash is collected:** "Capaian Kinerja Pengelolaan Sampah," Sistem Informasi Pengelolaan Sampah Nasional, accessed January 4, 2024, https://sipsn.menlhk.go.id/sipsn/.

267 **"I also see what's happening" . . . "toward the United States":** "Remarks Prior to a Meeting with Prime Minister Justin P.J. Trudeau of Canada and an Exchange with Reporters in London, United Kingdom," American Presidency Project, December 3, 2019, https://www.presidency.ucsb.edu/documents/remarks-prior-meeting-with-prime-minister-justin-pj-trudeau-canada-and-exchange-with-2.

268 **found five Asian . . . oceans from land:** Jenna R. Jambeck et al., "Plastic Waste Inputs from Land into the Ocean," *Science* 347, no. 6223 (February 2015): 768–71, https://doi.org/10.1126/science.1260352.

268 **"Everyone agrees" . . . "challenge is greatest":** "The Alliance Launches Today," Alliance to End Plastic Waste, January 15, 2019, https://endplasticwaste.org/en/news/the-alliance-launches-today.

269 **many were tiny . . . spend anyway:** Stephanie Baker, Matthew Campbell, and Patpicha

NOTES

Tanakasempipat, "Inside Big Plastic's Faltering $1.5 Billion Global Cleanup Effort," *Bloomberg*, December 22, 2022, https://www.bloomberg.com/features/2022-exxon-mobil-plastic-waste-cleanup-greenwashing/.

270 **even inspiring an *Onion* spoof:** "Officials Urge Americans to Sort Plastics, Glass into Separate Oceans," *The Onion*, February 16, 2015, https://www.theonion.com/officials-urge-americans-to-sort-plastics-glass-into-s-1819577493.

271 **An environment ministry official I meet . . . excessive packaging:** Novrizal Tahar, director of solid waste management, Indonesian Ministry of Environment and Forestry, interview with author, May 11, 2023.

271 **new study, published in 2020:** Kara Lavender Law et al., "The United States' Contribution of Plastic Waste to Land and Ocean," *Science Advances* 6, no. 44 (October 2020), https://doi.org/10.1126/sciadv.abd0288.

Epilogue

277 **"so many people that have to say yes, so many chances for it to die":** Alexandra Kahn, interview with author, August 16, 2023.

277 **A teenage activist . . . eighteen-meter scroll:** Dyson Chee, "An Introvert's Guide to Making Plastic History," OnlyOne, August 8, 2020, https://only.one/read/an-introverts-guide-to-making-plastic-history.

277 **"If you can organize" . . . "and counties":** Hawaii state Senator Chris Lee, interview with author, July 26, 2023.

278 **"If you don't sit down . . . invested in it":** Berkeley, California, councilwoman Sophie Hahn, interview with author, July 27, 2023.

279 **"It exists already" . . . "fair price":** Joan Marc Simon, Zero Waste Europe executive director, interview with author, March 21, 2024.

279 **$10 billion opportunity:** Thom Almeida and Llorenç Milà i Canals, "How 2025 Can Become a Tipping Point for Reusable Packaging Systems," World Economic Forum, January 21, 2025, https://www.weforum.org/stories/2025/01/tipping-point-year-for-reusable-packaging-systems/.

279 **"a better product . . . affordable as plastic":** As quoted in Beth Gardiner, "To Fight Plastic Waste, an Indonesian Campaign Aims High," *Yale Environment 360*, December 7, 2023, https://e360.yale.edu/features/tiza-mafira-interview.

280 **"there is less oil" . . . "the consumer":** As quoted in Gardiner, "To Fight Plastic Waste."

INDEX

Agung, Chresna, 259–60
Alajmi, Salman, 129
Alexander, Chloe, 154
Allen, Ben, 111, 275–76
Allen, Frederick, 18
Alliance to End Plastic Waste, 128, 268–69
al-Naimi, Ali, 115–16
Al Sharef, Fayez, 127, 131–32
American Chemistry Council, 69, 96, 141, 148, 150–51, 194, 195–96, 209–10, 268–69
American Legislative Exchange Council (ALEC), 203–10
American Plastics Council, 99, 100, 101–2, 141
American Progressive Bag Alliance, 196, 208–9
America's plastic boom, 57–82
    environmental regulation, 62-64, 68–71
    fracking and, 59, 65–68, 71–74, 76–77, 98, 132, 232–34
    health issues and, 59, 62–64, 77–82 (*See also* carcinogens; hormone-disrupting chemicals; microplastics)
    infrastructure, 57–62
    pipelines, 74–77

    pollution, 59, 62–64, 68–69, 81–82, 116–17, 164–65, 172–73
    racially segregated sacrifice zones, 77–78, 81–82
Amoco, 95, 97, 190, 192
Anderson, Gary, 83–85, 103–4
Antwerp, 237–46
Arabian American Oil Company (Aramco), 114–16. *See also* Saudi Arabian Oil Company (Saudi Aramco)
Arellano, Yvette, 57–62
Arisandi, Prigi, 247–51, 254

Baekeland, Leo, 14
Bakelite, 14
Banner, Jon, 221–22
Basel Action Network, 260–62
Beck, Nancy, 151
Belgium, 237–42
Bernays, Edward, 19
Bernstein, Roger, 93, 101
Biden administration, 63, 82, 149–50
bisphenol A (BPA), 139, 143, 145–46, 149, 152, 153
Blanock, Janice, 73
Blanock, Luke, 73
Bourque, Martin, 104–7, 112
Bower-Bjornson, Lois, 71–73

INDEX

Boxer, Barbara, 148
BP subsidiaries, 231–33, 234
Burkhardt, Delara, 220

Campen, Matthew, 158–59, 161
Canfin, Pascal, 220–21, 222, 224, 225
carcinogens, 40, 59, 64, 69–70, 73–74, 77–78, 80–82, 90, 144, 145
Carroll, William, 92
Celloplast, 187–89
celluloid, 13–14, 24, 83
Chemical Manufacturing Association, 141, 147, 150
Cheney, Dick, 171
Chevron, 67, 95, 97, 122, 150, 194, 223
Chik, 263–64
China, 32, 104–7, 123, 130–32, 249–50, 268
Chokola, Peter, 37–38, 39, 40
Citizen Coke (Elmore), 37
Coca-Cola, 37, 38, 39–40, 42–43, 47, 48, 52–53, 86, 87, 112
Colborn, Theodora, 133–37, 139–42, 144, 145, 146
Conkle, Jeremy, 178
Container Corporation of America, 83–84
Conway, Erik M., 22–23
Council for Solid Waste Solutions, 95–99
Crotty, Tom, 231, 236, 239, 243
"Crying Indian" commercial, 49–52
Cusack, Catherine, 55

Danzí, Maria Angela, 223–24
Darnay, Arsen, 45–46, 91
Davies, Fred, 113
De Croo, Alexander, 243–44, 245
DelBello, Alfred, 52
Dow, 41, 42–43, 95, 119–20, 126, 127, 150, 194, 223, 243–44, 268
Dubai conference, 124–30
Ducey, Doug, 199
Duer, Jacob, 128
Dumanoski, Dianne, 140–42
Dunaway, Finis, 50
DuPont, 15, 16–17, 18, 40, 42–43, 79–80, 95, 101–2, 147, 150, 151, 165, 194, 243–44
Duterte, Rodrigo, 252

Ecoton, 248–50, 256, 259, 265
Elmore, Bartow, 37, 54, 101
Enck, Judith, 276–77
endocrine-disrupting chemicals. *See* hormone-disrupting chemicals
Englebright, Steven, 190–92, 193–94, 210–12
Europe and the European Union
  container deposit laws, 55–56
  disposable packaging history, 41
  hormone-disrupting chemical regulation, 151–55
  Packaging and Packaging Waste Regulation development, 217–27
  plastic bag and single-use plastics ban, 213–17, 276–79
  producer-responsibility laws, 275–76
  waste management, 105, 107–10
Exxon and ExxonMobil, 22–26, 31, 67, 68, 95, 98, 108–10, 114, 116, 119, 150, 194, 223, 243–44, 268
extended producer responsibility. *See* producer-responsibility laws

Fadiyah, 254, 256–58, 259
Fawcett, Eric, 11–12, 14–16, 29
Formosa Plastics Group. *See Wilson vs. Formosa*
fracking, 59, 65–68, 71–74, 76–77, 98, 132, 232–34
Franklin-Wallis, Oliver, 85–86, 112
Freeman, Lew, 94–95, 97–98
Freinkel, Susan, 93, 188, 193

Galbraith, John Kenneth, 18–19
Gdula, Karen, 74–77
Geyer, Roland, 28–30, 280
Gibson, Reginald, 11–12, 14–16, 29
Gilmour, David, 240
Gone Tomorrow (Rogers), 51
Goorden, Thomas, 237–38, 242, 244–46
Gore, Al, 141
Gove, Michael, 154–55
Gramm, Phil, 170
Graves, Lisa, 206
Green, Adolphus, 35–36
Gunn, William, 45

336

INDEX

Hamblet, Trey, 66–68
Hamrick, Ronnie, 175–76, 184–85
Hansen, Karen, 182–83
Hendra, Lannawati, 253
history of plastic, 11–33
    consumerism and, 17–22
    current ubiquity of plastic, 26–30
    disposability incentives, 19, 21–22, 27, 29–30
    early plastic, 11–17
    early plastic applications, 12–15, 17–22
    reversal of supply and demand principle, 15–17
    tobacco industry's ties with oil industry, 23–26
history of selling plastics, 35–56
    beverage containers, 37–40, 42, 43, 45
    bottle bills, 47–48, 52–56, 107, 111
    food containers, 35–37, 42
    personal-responsibility framing, 43, 45–52
    pharmaceuticals, 41–42
    processed food, 43
    profit in manufacturing packaging, 40–42
    squeeze bottle, 41
    waste management warnings, 42–47
Hodgson, Paul, 20
Hollowood, John, 230–31
Honeycutt, Michael, 68–69, 82
Hopkins, Loren, 62–64
hormone-disrupting chemicals, 133–55
    background story, 133–37
    European Union regulation of, 151–55
    health dangers, 140–46
    plastics as source of, 137–39, 143
    regulation of, 146–51
Hoyt, Kenneth, 181
Hyatt, John Wesley, 13–15, 24

Imperial Chemical Industries, 11–12, 14–15, 30
India, 68, 132, 156, 165, 252, 263–65
Indonesia
    first signs of imported plastic waste, 247–51
    government's response to imported plastic waste, 252–58
    local bag bans, 279
    marine plastic pollution from, 268, 269–71
    production of waste, 262–67
INEOS, 229–46
    Antwerp plant construction, 242–46
    Antwerp plant permit process, 237–42
    background, 229–36
    Plastics Europe member, 223
    on recycling label system, 109–10
Innovene, 231–33

Jambeck, Jenna, 28–29, 269–71
Jellinek, Steven, 146–47, 148
Johnson, Amy, 177–78, 182
Johnson, Boris, 153
Jørgensen, Finn Arne, 55
Joseph, Stephen, 195, 210
Jurasek, Cheyenne (son), 175, 177–78
Jurasek, Dale, 172–75, 179–80

Kavanagh, John, 200–203
Keep America Beautiful, 48–52, 53, 100
Koch Industries, 205, 206–7
Kochland (Leonard), 205, 206
Krane, Jim, 116
Krause, Gerrit, 220, 221, 222, 225
Krishnan, Chinni, 263–64
Kuby, Lauren, 201–3

Law, Kara Lavender, 28–29, 271
Leonard, Christopher, 205, 206
Luján, Tatiana, 216, 227, 239–41
Lynch, Doug, 169

Mack, Walter, 38
Mafira, Tiza, 279–80
Mancusi, Peter, 53
Marfella, Raffaele, 157–58
Marius, Hugh, 46
McCluskey, Paul, 47
McCoy, J. W., 17–18

INDEX

McDonald's, 190, 191, 204, 220–23, 278
Meikle, Jeffrey, 21
Meloni, Giorgia, 223
Menefee, Christian, 70-71
Merchants of Doubt (Oreskes and Conway), 22–23
microplastics, 5–6, 25, 27–28, 90, 138, 155–61, 182, 211, 247, 249, 265
Mobil and Mobil Chemical, 42–43, 86–87, 88, 91, 93, 95, 97, 98–99, 114, 188, 189, 190, 191. *See also* ExxonMobil
Monbiot, George, 152, 155
Monsanto, 18, 20, 40
Moyers, Bill, 204, 205
Mudd, Sidney, 52–53
Muffett, Carroll, 22–26, 31, 33
Myers, John Peterson, 140–42, 144

Nabisco, 35–37
Nasser, Amin, 119, 121
neoprene, 79–82
Ní Chléirigh, Rose, 218–20, 226–27
Nixon, Richard, 47, 54
Novolex, 209, 212, 221

Oreskes, Naomi, 22–23
O'Sullivan, Grace, 217–18, 226
Our Stolen Future (Colborn, Meyers, and Dumanoski), 140–42
Outwater, Eric, 43, 90–91

Packaging and Packaging Waste Regulation (PPWR), 217–27
PepsiCo, 38, 40, 50, 52–53, 54, 86, 126, 128–29, 208, 223, 268
Perrin, Michael, 11–12
personal-responsibility framing, 2, 4–5, 9, 43, 45–46, 47–52, 86–87, 128
Peters, Frank, 36
phthalates, 138–39, 143, 152
Plastic (Freinkel), 93
plastic bags and single-use plastics, 187–212
    current ubiquity of, 27
    European Union bans, 213–17
    invention and widespread use of, 187–89
    lobbying against bans, 194–200, 203–10
    local bans, 190–203, 209–10, 211–12, 276–80
    state bans, 210–12
    state-level pushback against bans, 199–203, 208–10
Plastic China (documentary), 106–7
plastics industry
    background, 1–9 (*See also* history of plastic; history of selling plastics)
    bans on plastic bags (*See* plastic bags and single-use plastics)
    booming in America (*See* America's plastic boom)
    citizens standing up to (See *Wilson vs. Formosa*)
    defined, 6
    future directions for, 273–82
    global petrochemical web (*See* China; Europe and the European Union; Indonesia; Saudi Arabian Oil Company)
    guilt eraser (*See* recycling)
    health issues due to (*See* carcinogens; hormone-disrupting chemicals; microplastics)
    less-is-more principle, 280–82
    local bans, 190–203, 209–10, 211–12, 276–80
    private investment by billionaire (*See* INEOS)
    producer-responsibility principle, 275–77
    shifting blame and shipping waste (*See* personal-responsibility framing; waste management)
Plastics Industry Association, 94, 96, 126, 209
Pocan, Mark, 205
A Poison Like No Other (Simon), 156
pollution
    in America, 81–82, 116–17, 164–65, 172–73 (See also *Wilson vs. Formosa*)

338

INDEX

in Belgium, 238
global plastic pollution treaty, 3, 129
invisible poisons, 133–61 (*See also* carcinogens; hormone-disrupting chemicals; microplastics)
marine pollution, 213–17, 267–69
nitrogen crisis, 241–42
polyethylene, 12–13, 15, 16, 17, 21, 29–30, 41, 65–67, 89, 103, 122, 123, 131, 159, 170, 187–89, 214, 238
polyethylene terephthalate (PET), 40, 89, 99, 103, 107, 111, 131, 214
polypropylene, 17, 24–25, 65, 123, 170, 131, 234, 241
polystyrene foam (Styrofoam), 40–41, 57, 95, 97, 99, 190, 191, 192–93, 202, 222, 277
polyvinyl chloride (PVC), 15, 77, 138, 153, 166, 167, 169, 170, 176, 178, 179
Pratt, Larry, 204
Prattichizzo, Francesco, 156–58
Procter & Gamble, 6, 21, 42–43, 92, 110, 126, 204, 264, 268
producer-responsibility laws, 46, 275–77
Propaganda (Bernays), 19
Puckett, Jim, 260–62
PVC production, 166–67, 170. See also *Wilson vs. Formosa*

Ranganathan, C. K., 263–64
Ratcliffe, Jim "Sir Jim," 229–36, 238–39, 243–46. *See also* INEOS
Rebin, 254–56
recycling. *See* waste management
Resin Identification Code, 103–4, 111
R. J. Reynolds, 24–25, 205
Rockefeller, John D., 16, 114
Rogers, Heather, 51, 94
Rozenski, Phil, 209
Rozner, Van, 179

SABIC (Saudi Basic Industries Corp.), 16, 67, 120–22, 123, 124–27, 128–29, 183, 268
sachets, 260, 263–66

Sadara (petrochemical complex), 119–20, 127, 131
Salini, Massimiliano, 222–23
Saudi, Inc. (Wald), 114–15
Saudi Arabian Oil Company (Saudi Aramco), 16, 113–31. *See also* Arabian American Oil Company (Aramco)
background, 114–19
crude-to-chemicals transition, 119–24, 131–32
Gulf petrochemical conference, 124–30
Saudi Basic Industries Corp. (SABIC). *See* SABIC
Schmieder, Ronald, 188–89
Seanor, Bill, 188
Seeing Green (Dunaway), 50
selling plastic. *See* history of selling plastics
Setyorini, Daru, 247–52, 254–55, 257, 259, 260, 262–64, 265–67
Shell and Shell Chemicals, 3, 16, 24–26, 32, 33, 69, 71, 72, 74, 76–77, 150, 204, 207, 243–44, 268
shipping waste. *See* Indonesia; waste management
Simon, Matt, 156, 159
Single-Use Plastics Directive, 213–17. *See also* plastic bags and single-use plastics
Smith, Coy, 104
Smith, David, 198–200
Smith, J. Lucian, 39
Society of the Plastics Industry, 19, 41, 93, 94–98, 102–3, 104, 141, 190–91, 194, 209. *See also* Plastics Industry Association
Sonnenschein, Carlos, 138
Soto, Ana, 138
Sri Lanka, 183, 265, 268, 269
Starmer, Keir, 236, 245
Stefanelli, Leonard, 43–45
Stouffer, Lloyd, 41–42
Strand, Ginger, 50
Strasser, Susan, 100
Styrofoam (polystyrene foam), 40–41, 95, 97, 99, 190, 192–93, 202, 222

INDEX

Sullivan, Laura, 89
Surabaya Mekabox, 253
Susanti, Ariana, 266–67

Taylor, David, 268
Taylor, Robert, 79–82
Texas. *See* America's plastic boom; *Wilson vs. Formosa*
Thomas, Larry, 93, 102
Thulin, Raoul, 187
Thulin, Sten Gustaf, 187–89
tobacco industry, 23–26, 50, 204–5, 214, 222,
Toxic Substances Control Act, 146–49
trash. *See* waste management
Trisyanti, Dini, 266–67, 271
Trump, Donald, 267–68
Trump administration, 63–64, 82, 149–50, 151, 208

United Arab Emirates (UAE), 124–30

Vitale, Lisa, 180
von der Leyen, Ursula, 243–45

Wald, Ellen, 114–15
Wang Young-Ching "Y. C. Wang," 166–71. See also *Wilson vs. Formosa*
waste management, 83–112, 247–71
    background, 247–51
    chemical recycling, 108–10
    deposit bills (bottle bills), 47–48, 52–56, 107
    downside of disposability, 42–47
    marine plastic pollution, 213–17, 267–69
    personal-responsibility framing, 2, 4–5, 9, 43, 45–46, 47–52, 86–87
    propaganda and greenwashing, 48–51, 86–94, 98–101, 110–12, 128, 190, 192
    recycling, history of, 85–86, 90–91
    recycling, modern iteration of, 86–87
    recycling exported, 104–7, 252–62 (*See also* Indonesia)
    recycling labeling system, 102–4, 109–10, 111
    recycling lobbyists, 93, 94–98, 102–3
    recycling logo, 83–85, 103–4, 111
    recycling realities, 86–94, 101–4, 107–10
    research on, 269–71
Weyrich, Paul, 203–4
Wilson, Diane, 163–85, 243
Wilson vs. Formosa, 163–85
    case filing and trial, 177–81
    enforcement of, 184–85
    evidence against Formosa, 175–78
    Formosa employees' accusations, 172–75
    Formosa's background, 165, 166–70, 174–75
    ruling against Formosa, 181
    settlement and Wilson's trust, 181–83
    verdict's wide-reaching impact, 183–85
    Wilson's background, 163–65, 171–72
Woods, Darren, 68
Wyeth, Nathaniel, 40

Zhejiang Petroleum and Chemical, 123, 131

ABOUT THE AUTHOR

BETH GARDINER is an American journalist based in London. Her work has been published in outlets including *The New York Times, The Guardian, National Geographic, The Washington Post, Scientific American, Yale Environment 360,* and *HuffPost,* and she is a former longtime Associated Press reporter. Her first book, *Choked: Life and Breath in the Age of Air Pollution,* was named one of 2019's best by *The Guardian* and was a finalist for the National Association of Science Writers' Science in Society book award.